Was Darwin Wrong?

YES

By Richard B. Pittack, B.A., M. Div.

Was Darwin Wrong? Yes
By Richard B. Pittack

Published by Walden's Computer Services.

Printing History:
September, 2006. First edition.

Title: Was Darwin Wrong? YES
ID: 231595
ISBN: 978-0-6151-4124-4

Contents

APPENDIX

WAS DARWIN WRONG? - YES

The Evidence for Evolution Is *Not* Overwhelming

By

Richard B. Pittack, B.A., M. Divinity

A Commentary On The

Article in the November *National Geographic*, 2004,

"Was Darwin Wrong? - No"

"Do you not know?

Have you not heard?

The LORD is the everlasting God,

the Creator of the ends of the earth."

NIV

Isaiah 40:28

INTRODUCTION

Dear Readers,

At the age of fourteen, I became a creationist. I had believed in God from my earliest years and when I reached the age of eight, my mother enrolled me in a Bible course. Reading the prophecies recorded in the books of Daniel and Revelation, I understood that the history of Babylon, Medo-Persia, Greece, and Rome were predicted in the order of their rise and fall and long before the event. In this way, I became convinced of the Bible's authenticity. My convictions were further confirmed when attending the seminary. I learned the many ways for establishing that the book of Daniel was not only prophetic in content but was actually written in the 6th century B.C. before the rise and fall of the nations just mentioned. In accepting the Bible as the Word of God, its opening account recorded in the book of Genesis - God's marvels of creation - became the main focus for my biblical studies. I had always, even before my convictions concerning God's word, accepted the creative acts of God but it wasn't until further along in my life, while walking through the woods in Pennsylvania and observing nature's scenery, I was convinced in my heart and mind that such beauty could only be explained as coming forth from the hands of the Creator.

The following will give you some idea as to how I reasoned as a young man. I began to let my imagination run wild. I soon discovered that I couldn't think of anything that doesn't have existence. Try it! ...Can you think of anything fantastic though it might be that hasn't been thought of before? Can you picture any thing in your mind that hasn't been imaged before? I then thought of the wonderful and complex articles in nature and the universe as evolving from nothing; with nothing to guide thought patterns; with not even some mind (any mind) to create something(any thing)from nothing. From that point until this day, being sixty-nine years old, I have accepted all creational designs fashioned after His patterns. The Creator Himself generated my faith; even faith is a gift from Him. The person who believes that life was spontaneously initiated and who believes that all nature was fashioned by chance and out of nothing must have a self-generated faith; how can it be otherwise? I did not know about the following statement by the British philosopher, G.K. Chesterton. However, as a creationist, I think of how it applied to my situation of many years ago and why I could not have become an evolutionist: "It is absurd for the evolutionist to complain that it is unthinkable for an admittedly unthinkable God to make everything out of nothing, and then pretend that is more thinkable that nothing should turn itself into everything." As a young man, my faith in God was evoked and throughout my life, the more I studied and prayed, the more I had my faith strengthened and confirmed.

I wrote the following work for my family and friends and all those interested in the subject of evolution and creation. I have a difficult time conversing following my stroke but I am still able to place my thoughts on paper. Some of the readers of my book will be young people who have been educated in the public school system. The public schools teach only one viewpoint – evolution. What I have written may appear new to you but I trust that these words, recorded by the pen of a creationist, will weigh in on the thinking of those of you who are of the Darwinian persuasion.

Should you have the *National Geographic* from November of 2004, it would help you to follow my reasoning. However, it is not absolutely necessary. I have placed David Quammen's arguments in CAPITAL LETTERS.

In November of last year, 2004, I read the article "Was Darwin Wrong?—No." It was an annoyance to me in realizing how much of the public was being hood-winked into accepting such unsound concepts. Upon reading Quammen's so- called scientific observations, I began making short comments on the pages of his article. These notes became more copious until I finally decided to compile them into book form. I wrote for three weeks, often all night, in placing this work together. In the past five months, parts of the contents have been reworked with three added appendixes.

I have intentionally refrained from adding Biblical scripture except in those instances I judged it to offer additional meaning to the context or felt compelled to draw a contrast between my belief and the evolutionist's belief. Otherwise, I have attempted to make the case for creation based only on the facts of science. In this work, the quotes are taken from the books and articles of qualified scientists and other professionals.

Each section is geared to demonstrate that speciation, by natural selection, has been directly observed in nature. However, although microevolution occurs, macroevolution is not demonstrable within nature nor has it been validated by one single empirical discovery since Darwin's <u>Origin</u> of 1859. For this reason creationism cannot be "stamped out" as a competing theory; it must be coped with as a rival.

Nicky Perlas comments:

"The neo-Darwinian theory of evolution is not only suffering from an identity crisis but may also be radically transformed to account for the growing number of scientific anomalies that continue to plaque it…the symposium was convened under the auspices of the 148th Annual Meeting of the prestigious [AAAS] held from January 3, 1982 to January 8,1982 at Washington, D.C.

The symposium was a disappointment to the true believers of Neo-Darwinism. Implicit in their counter-offensive to stamp out creationism was the recognition that they had to contain and mend the fissures that were increasingly undermining the scientific foundation of their own neo-Darwinist position. To their dismay; the symposium aggravated and deepened the fissures."

Towards, vol.2 (Spring 1982)

"Neo-Darwinism Challenged at AAAS Annual Meeting"

Pp.29-31

May this present work confirm to its readers that Darwinism cannot, in any of its forms, justifiably serve as a "counter-offensive" to creationism!

I dedicate my words and thoughts to my dear mother who departed from this world in November of 2003, at the age of 92, and who silently awaits the appearance of her Creator whom she then will see face to face. I also offer to my dad, who is presently 96 years old, my love and respect. I thank you dad, for taking me to the museum of The History of Natural Science, the Aquarium, the Planetarium, the Zoological garden, and to scientific lectures in the city of Philadelphia. But mainly, for providing me with the type of family background that taught me how to combine my love for the natural sciences with my knowledge of the love of God and His created works.

Respectfully,

Richard B. Pittack

Palmdale, California,

November 21, 2005

CHAPTER ONE

THE UGLY BUT BEAUTIFUL
MOLE RAT

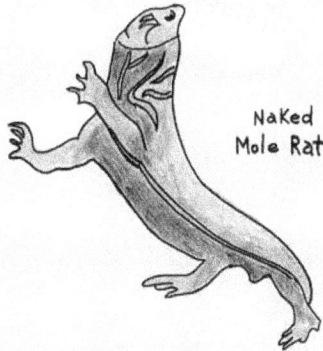

Naked
Mole Rat

5   RBP

THE NAKED MOLE RAT SHOWS
THAT MAMMALS CAN EVOLVE.
THE EVIDENCE FOR EVOLUTION
IS OVERWHELMING.

National Geographic, Pp.4-5

Comments:-David Quammen cleverly starts out his article on the evidence for evolution by alluding to the naked mole rat; a mammal which by his appearance, is supposed to show that mammals can evolve and also to draw our attention away from the complexity of animal structures and design in nature.

Quammen, of course, does not start out with a mammal-model of the duck-billed platypus which is a bit problematic when attempting to establish the reliability of evolution's credo. The duck-billed platypus is a creature which Quammen and other "evolutionists wish never existed" (Gish). This creature is a mammal, and yet it has a duck-bill, webbed feet, and lays eggs. But for now, we will focus on the mole rat.

When thinking of the mole rat we should think of biological adaptation. The mole rat is a burrowing rodent, superficially resembling a mole. Unlike the mole which is not found in Africa, the mole rat has three families living in Africa. What is more, the mole is classified in the Order of Insectivora and feeds on invertebrates, while the mole rat is classified in the Order of Rodentia and eats roots and tubers. The mole rat is perfectly adapted to his environment but the environment can only determine the survival of the fittest and not the arrival of the fittest – a principle that will be understood afterward when natural selection is discussed more fully in coming chapters.

I will admit that this creature is rather unsightly but, after all, beauty is in the eye of the beholder and to Mrs. Mole rat, Mr. Mole rat appears to be rather dapper. She rather likes his unassuming and modest appearance. She would agree with Ecclesiastes 3:11 "He hath made everything beautiful in his time" – including the mole rat.

David Quammen also likes the simplicity of the mole rat's appearance. That is why he wrote, "The naked mole rat shows that mammals can evolve." But how does Quammen arrive at this suggestion so easily? Is he justified in concluding that because a creature appears to be simply constructed, that it is necessarily a product of transmutation? The mole rat did not evolve from a totally different animal of the past. He was born a mole rat, like every mole rat from the beginning of time, because his parents were equipped with the proper genes and chromosomes for maintaining the perpetuation of their species. Outside the mating of his parents there is no logical reason for the birth of the mole rat. The fossil record is not a progressive "record" of the evolutionary past of the mole rat. There are no transitional forms leading up to this mammal under consideration. Paleontology cannot support the birth of the mole rat for it has no "birth certificate" in its files. That is, nothing in the records to account for its origin.

There are three families of mole rats living in Africa. Natural selection has worked on the genetic information already in place and thus, the gene pool of the mole rat allowed a selection of smaller sets of genes to produce the differences in the three families of mole rats. No new genetic material could have been added to create another type of animal. The three families arose not by the introduction of anything new into the gene pool but by a simple recombination of existing genes. Recombination is defined as new combinations of traits different from those exhibited by the parents.

As we shall see, in further sections, not only is the evidence for evolution not overwhelming; it is based on pure speculation and fantasy. Michael Denton claims:

"That it [evolution] is neither fully plausible, nor comprehensive, is deeply troubling. One might have expected that a theory of such cardinal importance, a theory that literally changed the world, would have been something more than metaphysics, something more than a myth."

"Evolution: A Theory in Crisis"

P.358

Most of the tributaries of knowledge that evolutionists turn to in their quest to understand nature are based on valid scientific facts but it is the wrong interpretation of those facts and complete disregard for others, that lead men to adhere to the vain philosophy and false conceptualism of evolution. Other tributaries of knowledge such as recent findings of creation scientists are completely rejected because of prejudice and, it should be added, at a great loss to the scientific world.

CHAPTER TWO

CLASSIFIED BARNACLES
AND THE TREE OF LIFE

The Tree Of Life

In the *National Geographic*, November 2004, page 6, David Quammen offers this character build-up for Charles Darwin:

DARWIN WAS A SHY, CONSERVATIVE MAN,

WHO ASKED PENETRATING QUESTIONS.

Comments:-I can think of one penetrating question that Darwin never answered - What is the origin of any new species of plant or animal? In his book about origins, he never gave a specific example. He discussed variation within the species but never attempted to prove vertical evolution. I guess he was too shy.

Some Creationists, in a bantering or joking way, make the claim that Darwin never spoke about origins in his book on origins. But this is an over-sight. Darwin did mention origins coupled with his candid remarks, "We have no facts to guide us."

Darwin, with some hesitation wrote these words in *The Origin of Species*, P. 96:

"Looking to the first dawn of Life, when all organic beings, as we may believe, presented the simplest structure, how, it has been asked, could the first steps in the advancement or differentiation of parts has arisen?... ...But as we have no facts to guide us, speculation on the subject is almost useless. It is, however, an error to suppose that there would be no struggle for existence, and, consequently, no natural selection, until many forms has been produced ... ... But, as I remarked towards the close of the introduction, no one ought to feel surprise at much remaining as yet unexplained on the origin of species, if we make due allowance for our profound ignorance on the mutual relations of the inhabitants of the word at the present time, and still more so during past ages."

One can see that Darwin did not have much to go on when he sketched his much loved metaphor, a tree of life (P.8 of the *National Geographic*).

IN AN 1837 NOTEBOOK DARWIN SKETCHED HIS FAVORITE
METAPHOR: A TREE OF LIFE, ITS TWIGS AS SPECIES. THEN,
BELIEVING NO ONE SHOULD SPECULATE ABOUT SPECIES
"WHO HAS NOT MINUTELY DESCRIBED MANY," HE SPENT
EIGHT YEARS, CLASSIFYING BARNACLES. BY 1854 HE WAS
KNOWN AS A BARNACLE EXPERT---THOUGH NOT YET AN
EVOLUTIONIST.

"Then, believing no one should speculate about species ..." writes David Quammen. This statement is extremely odd when the whole metaphor, the tree of life, is built on nothing but speculation. The ignorance that went into the making of this tree already has been confessed by Darwin. This tree-metaphor will be shortly exposed for what it is---an illustration embracing Darwin's deceptive evolutionary philosophy.

There are, at least, four admissions in Darwin's statements:

*We don't have any facts to guide us in the first steps on how animal parts differ.

*We have no explanation on how species came into existence.

*We, at present, have a profound ignorance of how species interrelate.

*We have a more profound ignorance, when it comes to the past, of how species interrelate.

Quammen informs us that Darwin spent eight years classifying barnacles and that made Darwin a barnacle expert. The reasons for Darwin becoming a barnacle expert were twofold: that Darwin would become, after eight years of collecting barnacles, a person who is automatically free of speculation when it comes to species --- all species; the implication being, of course, if you aren't a barnacle expert, you have forfeited your right to even think about species and the tree of life. Not being a barnacle expert takes away your powers to reason and automatically makes you bound to Darwin's interpretation of transmutation of species. How strange! – the study of one particular species, after eight years, transformed the naturalist Darwin into the expert evolutionist who knew about all species which existed on the face of the earth for so-called millions of years.

We will now consider the tree of life and Darwin's speculation concerning it. The "Tree Of Life" is, actually, a well thought-out metaphor Darwin describes in his *Origin*, P. 99. He pictures the chain of life in a tree-like configuration, dividing and re-dividing into myriads of branches. In Darwin's simile, each branch connects (?) with the trunk and, thus, allegedly demonstrating that all animals come from a common source. Darwin's simile was a well thought-out metaphor but—COMPLETELY INCORRECT. The branches, in fact, do not connect with the trunk. The only real fact about Darwin's tree of life is that the tips of the branches represent the species of living organisms. Everything in the tree metaphor is hypothetical. Indeed, the discoveries of modern science have dispelled the notion that the root of the tree is the common ancestor that gives rise to all other life depicted on the tree. Also, the branching lineages have been disproved by the facts contained within the fossil record. The majority of today's qualified scientists recognize Darwin's process of modification, or the differences among organisms, as applying only to genera and not capable of producing new orders and classes. Therefore, the tree is not complete; the animals, through the course of time, are not branching off from a common ancestor; the fossil record indicates no such thing; there are major groups of animals that have no transitional species between them. The main proof of evolution - the paleontological record - has missing links with its vital information absent. This is why Darwin wrote, "...we make due allowance for our profound ignorance on the mutual relations of the inhabitants of the world at the present time and still more so during past ages." Meaning, there is no evidence for origin of the species in real life (and its association with the trunk of the tree) and less evidence in dead life - the fossil record - (and its association with the trunk of the tree).

It was a painful experience for me as I traced Darwin's "Tree Of Life" through pages 86 to 93 in the *Origin*. From the pen of one who was supposed to be free of speculation, came forth speculation after speculation, conjecture after more conjecture, and theorizing on top of theorizing. At each turn in Darwin's explanation, he had a series of dotted lines indicating that there is no connection of the twigs and branches with the main trunk of the tree. The word "supposed" is used frequently in his explanation [Darwin, in the Origin, uses the phrase "we may well suppose" 800 times to explain things that are not true]. For examples:

* "a sufficient amount of variation is supposed to have accumulated…"

* "is supposed in the diagram to have produced variety…"

* "is supposed to have produced two varieties…" etc, etc

Let us face it! The tree of life is really a shameful model for evolution.

Darwin, in summary, writes:

"As buds give rise by growth to fresh buds, and these, if vigorous, branch out and overtop on all sides many a feebler branch, so by generation I believe it has been with the great Tree of Life, which fills with its dead and broken branches the crust of the earth, and covers the surface with its ever-branching and beautiful ramifications." P.100

And so, with these eloquent thoughts but lacking substance, Darwin brings this part of his theory to a close. Mark Twain said, "Eloquence is the essential thing … not information." The tree of life is also a tree of death. The fossil record of dead animals, nevertheless, speaks more eloquently than Darwin. Its plain testimony of no transitional species breaks off the twigs and branches from the main trunk in Darwin's clever but pointless evolutionary tree. Darwin's dotted or dashed lines indicate that the "supposed" evolutionary connections to the main trunk were only hypothetical and unsupported by the fossil record. The tree was held together by theory alone and that is why it has toppled over in the woods of debauched ideas, not only for creationists but for many evolutionists as well. In fact, the tree has not only toppled over but it has been "uprooted." Jonathan Wells who holds Ph.D.s from both Yale University and the University of California, Berkley has written in his must-read book *Icons of Evolution Science or Myth?* Page 31:

"Ten years ago it was hoped that molecular evidence might save the tree, but recent discoveries have dashed that hope. Although you would not learn it from reading biology textbooks, Darwin's tree of life has been uprooted."

Barnacles:

Added Comments:-Darwin collected barnacles for eight years. As he ventured down to the sea he was eager to collect these marine crustaceans, these complex invertebrates, to prove his case for transmutation. Darwin knew that fossilized barnacles were found in the low levels of the Cambrian rocks. I wonder how he explained the Cambrian feature of complex animals whose only ancestors were bacteria and blue-green algae! I wonder if he reasoned, since barnacles were able to be traced all through the fossil record until present days, this might be an argument against his belief in progressive transmutation! This was the least of the problems that confronted Darwin's thoughts. Not only were traces of the fossilized barnacle found in the Cambrian level of rocks but the highest levels of the biological hierarchy appeared right at the beginning of the geological column. Darwin was well aware that the

Cambrian level started out with completely developed and complex organisms equal to their modern counterparts. Many of these forms were extinct but all followed a natural classification unit. That is, there were phylum-level differences right from the start

This phenomenon in nature was later to prove fatal to Darwin's evolutionary theory. How could evolution have occurred in worms, crustaceans, or brachiopods without any precursors? How could modification work in a series of slow steps and organisms appear so fully developed without having remains of their ancestors demonstrating such development? No small wonder that Richard Dawkins wrote on page 229 of his book, *The Blind Watchmaker* – "…the Cambrian strata of rocks … are the oldest in which we find most of the major invertebrate groups. And we find many of them already in an advanced state of evolution, the very first time they appear. It is as though they were just planted there, without any evolutionary history. Needless to say, this appearance of sudden planting has delighted creationists." In response to Dawkins…
… …

Creationists do not delight in the fact their argument against a common evolutionary ancestor in the tree of life is <u>supported</u> but rather they delight in the fact that their understanding of factual science is <u>confirmed</u>. Science is not a game to be won by the side that can set forth the best arguments. Rather, it is a tool to assist men in their search for truth in the natural world. The search for truth is serious business and science is not an entertainment specialty designed for those individuals who love to engage in heated debates. Science has a greater purpose than to glorify those individuals who are the most skilled and proficient in expressing either the philosophy of creationism or the philosophy of evolution.

Returning to the crucial problem that Darwin faced in the sudden appearance of fossil remains already in an allegedly advanced state of evolution, he wrote in *Origin of Species* – "To the question why we do not find rich fossiliferous deposits belonging to these assumed earliest periods prior to the Cambrian system, I can give no satisfactory answer …. The case at present must remain inexplicable; and may be truly urged as a valid argument against the views here entertained." P.253

Darwin could not account for the force of the Cambrian level but he tried. He used Lyell's metaphor of likening the geological record to a volume recording the "history of the world imperfectly kept." "Of this volume, only here and there a short chapter has been preserved; and of each page, only here and there a few lines. Each word of the slowly-changing language, more or less different in the successive chapters, may represent the forms of life, which are entombed in our consecutive formations, and which falsely appear to have been abruptly introduced. On this view, the difficulties above discussed are greatly diminished, or even disappear."

Origin of Species

P.255

What a strange conclusion! Darwin claims that the imperfection of the geological record removes the difficulty of the problem which concerns the transitional forms of life. In fact, the problem disappears. And what stands in place of the imperfection of the geological record? – The establishment of the theory of the transitional forms of life. Was it that the forms of life have made their false appearance in the Cambrian level or was it that Darwin's reasoning was faulty? There are, in fact, no transitional forms leading up to the Cambrian level. This is one disappearing act that Darwin could not account for with the magic of evolution. So much for his candid (?) remark "The case at present must remain inexplicable." Not only did he make an attempt to explain the problem but he said his theory was established through his explanation.

On page 8 of the *National Geographic,* just above the picture of the hand sketch of the "tree of life", can be found the words of David Quammen, "NO ONE NEEDS TO, AND NO ONE SHOULD, ACCEPT EVOLUTION MERELY AS A MATTER OF FAITH." It took a matter of faith for Darwin to see his unconnected twigs and branches as being connected to the tree of life; a matter of faith to acknowledge the problem of transitional forms as being "inexplicable" only to come up with an explanation (?) in spite of the insurmountable circumstances; a matter of faith to believe that his theory of transitional forms was established regardless of what the empirical evidence of the geologic record indicated; a matter of faith to see the first, original, evolutionary ancestor (?) as arising spontaneously. When it comes to origins and the dogma of transmutations, the evolutionist must evoke faith - he has no other choice.

Evolutionists assert that life did not arise from the hand of God but from organic antecedents. They recognize, as did Darwin, the imperfections of geologic evidence but they say you can't get around the biological supposition of the reality of life in the pre-Cambrian level. Evolutionists note that all the major phyla excepting the chordates are found in the

Cambrian strata. With this in mind, they look for the evolution of these phyla before the time of the Cambrian. However, very few organisms appear beneath the Cambrian level and there is no evidence of ancestral forms leading to the complexity of life found in the Cambrian strata. Most pre-Cambrian finds are inorganic structures, calcareous algae, and algal and fungal filaments. The fact remains that very little has been discovered below the Cambrian in spite of the evolutionary supposition of life's beginnings in the pre-Cambrian level.

Jonathan Wells writes:

"...so our present understanding of Precambrian history is far better than Darwin's...

"Many paleontologists are now convinced that the major groups of animals really *did* appear abruptly in the early Cambrian. The fossil evidence is so strong, and the event so dramatic, that it has become known as 'the Cambrian explosion or biology's big bang.'"

Icons of Evolution Science or Myth

P.37

A few evolutionists charge creationists with citing the argument that God selected the base of the Cambrian because of its initial complexity for His days of creation. Most creationists, who are not <u>progressive</u> creationists, would never argue in this manner. On the one hand, our earth is under the curse of sin because of the transgressions of wicked men. Presently, beneath our feet, are the fossilized remains of not only mankind but animals and plants, as well. On the other hand, when God first spoke this planet into existence, He pronounced upon it His blessings of "good" and "very good." This was prior to the form of death rearing its ugly head above the earth. Because mankind rejected the true Creator and went forth seeking false gods, God said, "My Spirit will not contend with man forever." Because the earth was corrupt and filled with violence, God caused the waters of the Great Flood to flow from springs of the great deep and the floodgates of heaven. The creation came first and then sin with the fall of man; finally God's judgments in the form of a devastating Flood. The catastrophic results of this judgment are evidenced in the rocks and sediments of the earth and the fossils bear testimony of the unnatural forces that once prevailed upon it.

The evidence for the Great Flood is so overwhelming that the evolutionary tenet of uniformity is only a delusion. [Uniformity is the assumption that the present geological agencies are sufficient to explain the record of past history in the rocks] A few examples of evidence for the biblical Flood will now be given for your consideration: the cross-bedded sandstone in Monument Valley and Zion National Park, Utah. The sedimentary rock shows evidence of having been washed into place by currents with such volume, that natural results must give way to a Universal Flood. Coal beds contain upright trees that have been washed into place with strong currents. In some instances, these trees go through several layers of coal and intervening sandstones. Sometimes, the trees are head downward showing that the material was not collected in a natural manner. In the Colorado Plateau there are cross-bedded shales and sandstones from ten to a hundred feet thick for an area of 100,000 square miles. Any attempt for attributing these phenomena to uniformity is an attempt to circumvent the more logical issue of Diluvialism.

Harold Coffin makes this crucial observation:

"As we dig into the rocks and look at the fossils buried in these uplifted sediments we come away convinced that no gradual development from simple to complex has occurred in the history of life on the earth. An intelligent Creator filled the seas with swarms of living creatures of many diverse types. When it became necessary for Him to remove wicked men from the earth by a flood, many of the creatures of the seas were destroyed. They left no traces of their ancestors, because they had none, and in many cases they left no descendants, because none survived the Flood to perpetuate the species."

Creation – Accident or Design?

P.173

In answer to the charge above, God did not select the base of the Cambrian for building blocks during His creation week. For one thing, sin had not yet entered the world. Sin brings death and fossils are the remains of dead animals and plants. For another thing, the flood was future. God's Spirit was not hovering over this planet with its various systems for there were no large sedimentary systems. The Cambrian level did not exist with its broad range of fossils. However, in Darwin's day, it did exist and he not only had the historical, geological problem of fossils to contend with but he also had the geological, physical properties to explain solely on the basis of uniformity. This latter problem would become forever difficult since the forces which deposited the fossiliferous sediments were supernatural and inexplicable to the materialistic mind-set.

When Darwin appears to be self-scrutinizing in setting forth the arguments against his theory, he still deems his response to these arguments as efficient and convincing. In other words, his supposed honesty revealed in his admissions actually gives us insight into his super ego. He continued to vaunt his evolutionary theory by believing his rather simple and effortless explanations would convince his readers. But through the passing of time, the folly of his explanations became apparent. The imperfection of the fossil record, on a global scale, was his number one defense. He claimed that fossil collecting was a young science; the existing collections of fossils were far too inadequate to give examples of the truth of the continuous array of life's forms; only a small percentage of fossil-bearing strata had been explored and there was not, at this present time, sufficient evidence to establish his theory. Conversely, fossil collecting has reached every continent. There are millions of fossils available for scientific study. An adequate sampling of life's full riches has been made.

Darwin also had to account for the local issue of why we do not find a "graduated organic chain" or evidence for transitional species in continuous rock strata. He argued that strata may seem to be continuous but they are not. Sediment accumulates only during gentle subsidence of basins and not only that but transformation does not take place for any particular organism because inhabitants are continuously changing and organisms will track their moving living quarters. These two reasons are supposed to explain why there is lack of evidence for the transmutation of species. Darwin attempted to keep his theory of evolution intact no matter what were the prevailing circumstances. He reasoned that the sequence of deposition of the stratified beds was purported to demonstrate evolution through gradual development of organic life forms. However, if his theory was not demonstrable in the fossil record it was because of missing strata or organisms tracking their environment. Darwin called "heads" as he flipped his evolutionary coin into the air but his theory could not lose – his coin was two-headed. Militating against his circular reasoning is the fact that there is not a complete geologic column discovered at any one place in the earth that demonstrates exact sequential order of fossils. This makes Darwin's theory impossible to prove one way or the other. In addition, the supposed missing systems of strata are without any physiological evidence. For the most part, there is *conformity* between levels of stratification. Furthermore, rather than organisms tracking their environment, it is more likely that these organisms were either washed into place due to various water currents from other areas or were buried during the Great Flood in their natural environment or zones.

In reading Darwin's defense issues in *Origin's* 10th and 11th Chapters (Pp.234-277) – "On The Imperfections Of The Geological Record" and "On The Geological Succession Of Organic Beings" it is not easy to see how any modern geologist would conclude that Darwin argued convincingly and decisively in defending his theory involving the "finely graduated organic chain" or gradualism in terms of natural selection. Having developed weak argumentation for the problems of why we do not discover intermediate fossil links on a global and local scale, he now attempts to explain away sudden mass extinction that appear throughout the geologic column. Darwin wrote about "the old notion of the inhabitants of the earth having been swept away by catastrophes at successive periods is very generally given up ..." Darwin was referring to the major cause for extinction views of Baron Cuvier who, at the beginning of the nineteenth century, concluded that there had been extensive world catastrophes which accounted for the extinction of created basic types. This view had been abandoned by most geologists of Darwin's day and by all geologists of the present day. Cuvier's theory of multiple floods, of which the Noachian Deluge was the most recent, is called catastrophism. This is not to be confused with the Universal Flood which was one catastrophic event described in the book of Genesis. Cuvier's theory did not withstand the test of time but the Universal Flood still continues in opposition to the doctrine of uniformitarianism – "the present is the key to the past." Darwin attempted to reason from the standpoint of uniformity to account for the extinction of species which was often deemed a "mystery" of the natural world. Modern evolutionary geologists, very reluctantly, have returned to catastrophism in the quest of answering the questions in geology that uniformity fails to answer. I say reluctantly because they do not want to be confused with creationists who believe in the Noachian Flood.

Uniformity does not provide the convincing answers for the conviction that geology can be understood in terms of slow processes acting over long time periods. Approximately ten years ago, while working as an x-ray technologist, I met a young man who was studying to become an industrial geologist. He was carrying a large text book on the geology of southern California. I engaged him in conversation and the exchange went something like this:

Richard – "You have quite a large text book."

Student – "Yes sir! Geology is a complicated subject."

R. - "Do you mind if I look through your book?"

S. - "Go for it!"

R. - "I have been studying this subject for over forty years."

S. – "That's sixteen years longer than the date of my birth."

R. – (After perusal of the book) "I notice every Geologic Era, the Paleozoic, the Mesozoic, and right on up to the Cenozoic, that these time divisions were filled with catastrophic events."

S. – "You are absolutely correct."

R.– "What does this do for the doctrine of uniformitarianism?" (Expecting a lengthy answer).

S. – "Not much, sir."

R. – "Don't geologists study how to refute catastrophism?"

S. – "What for? You can't deny the obvious."

R. -"It sounds to me like you're saying geological events cannot be accounted for by the doctrine of uniformity.

S. – "Uniformity cannot always be relied upon for the best response to problems because it can not be applied to most geological events. Even in those rare cases of application, the uniformity principle can only be assumed."

R. – "Your response has confirmed what I have always believed about uniformity. You have certainly given me food for thought and I do thank you."

Within this section on "Barnacles" a number of catastrophic events have been mentioned in connection with the Universal Flood. Because Darwin, in defense of his theory, continued to focus on uniformity and to shun catastrophism as the major cause for extinction, two additional catastrophic events will be presented: It takes 10 feet of forest vegetation to produce one foot of coal. The theory of uniformity cannot possibly account for some coal seams that are 30 to 400 feet thick. The theory that coal formations were the result of the slow accumulations of peat in bogs is not suitable for explaining the facts of nature. A better and more reasonable explanation is the Universal Flood, which formed rafts of trees and other plants and washed them into large accumulations. These accumulations of organic materials became waterlogged and, covered with layers of sediments, hardened into coal. Also, the thin sedimentary units in a cyclic rhythm of deposition between such accumulations cannot be explained on the basis of uniformitarianism; the accumulation of fish fossils is unheard of in the present natural world. It doesn't take but a short time for any dead fish remains to be entirely dissipated. Present day fossilized fish are difficult to find except under rare circumstances. And yet, there are myriads upon myriads of fishes found throughout the Devonian that were simultaneously killed and were deposited in the sediments only a short time after their death. Their flesh, liver, and alimentary canal were intact before they were sealed by the Universal Flood. Modern experiments demonstrate that, in a matter of days, dead fishes begin to lose scales and are attacked by crabs and other fishes which begin their nibbling process. Fish fossilization calls for rapid entombment and finding thousands and thousands of these individuals, renders the doctrine of uniformitarianism null and void.

In the middle of the 19th century, Darwin sat in his comfortable study penning out a defense for his theory on the transmutation of species. In part, he defended his theory by pointing to historical geology interpreted in the light of uniformitarianism. Darwin, at the time of his writing the *Origin*, little realized that he soon would become the focus of a prophecy made in 66 A.D. by "Simon Peter, a servant and apostle of Jesus Christ." Peter was in a Roman prison and soon would entreat his executioners that he might be nailed to the cross with his head downward. He deemed he was not worthy to die in the same way his Master had died thirty-five years previously. Peter believed in prophecy and, in the Spirit, made a prophecy of his own concerning conditions that would exist in the world prior to the second coming of Christ. With prophetic eye, Peter saw Darwin and other scientists who rejected the Universal Flood and favored the doctrine of uniformity as the key which opened up past geological events.

Peter wrote in his second letter, chapter three, and verses 1-9: "Dear friends, this is now my second letter to you. I have written both of them as reminders to stimulate you to wholesome thinking. 2 I want you to recall the words spoken in the past by the holy prophets and the command given by our Lord and Savior through your apostles. 3 First of all, you must understand that in the last days scoffers will come, scoffing and following their own evil desires. 4 They will say, "Where is this 'coming' he promised? Ever since our fathers died, everything goes on as it has since the beginning of creation." 5 But they deliberately forget that long ago by God's word the heavens existed and the earth was formed out of water and by water. 6 By these waters also the world of that time was deluged and destroyed. 7 By the same word the present heavens and earth are reserved for fire, being kept for the Day of Judgment and destruction of ungodly men. 8 But do not forget this one thing, dear friends: With the Lord a day is like a thousand years, and a thousand years is like a day. 9 The Lord is not slow in keeping his promise, as some understand slowness. He is patient with you, not wanting anyone to perish, but everyone to come to repentance." NIV

Peter wrote specifically to the future skepticism of Charles Lyell, founder of the modern evolutionary theories of geology and to Charles Darwin, founder of natural selection as the mechanistic force behind evolution of the species. Both these men firmly established the evolutionism now prevailing throughout the world. Both would maintain that uniformitarianism and not catastrophism by a Universal Flood, was the answer to the past history of nature recorded in the fossils of the geological column. Darwin, in writing his *Origin*, didn't pay the least bit of attention to Peter's words mainly because they came straight from an epistle which had been canonized as a part of the Bible. Peter mentioned Creation of the World and the Universal Flood – two events that could never be accounted for by Darwin's uniformitarianism and which events provided far better explanations for the facts seen in nature, than evolution and uniformity could ever hope to provide. Thus, Peter's New Testament letter not only supports the geographical universality of the Great Flood but condemns the doctrine of Uniformitarianism.

Again, catastrophism on a major scale must be recognized as the only possible rationalization for species extinction that has been noted in the geological column. Darwin *was wrong* in his rejection of the Biblical Flood as a key factor in the interpretation of geological events. Rather he sought to uphold geological events as though they were the result of deep time. He used the false key of uniformitarianism (the assumption that the present agencies are competent to explain all the past) - the very thing that the apostle Peter wrote against.

The Cambrian level of advanced species *could not serve* as a testimonial supporting evolution, geared to fit Darwin's philosophical thinking of uniformitarianism. Nevertheless, it *definitely could serve* as confirmation of creationism, to fit the philosophical thinking of those who believed in the biblical Flood. The one philosophy (Evolution) is contrary to the facts of nature; the other philosophy (Creation) is in harmony with the facts of nature. Evolutionism is a self-generated philosophy based on a presumptuous faith; Creationism is a God-generated philosophy based on a faith imposed by the facts of nature and inspired by the reality of a

Creator. Readers, where do you think we should go for our facts regarding natural history or the history of recorded events in this world's geology? —to the light of *biblical Revelation* or to the darkness of *The Origin of Species!*

Natural history is a man-made term. It implies that God or miracles do not have a place in earth's history – all things happened naturally without divine intervention. We have read the words of the apostle Peter who, in prophecy, spoke to this very way of thinking – "Everything goes on as it has since the beginning of creation" (The motto of Scoffers, signifying their stance on uniformitarianism and their materialistic attitude regarding *natural* history). Peter goes on to write – "But they deliberately forget that long ago by God's word the heavens existed and the earth was formed out of water and by water. By these waters also the world of that time was deluged and destroyed."

The Great Flood was a very unnatural event in the natural order of creation. But the flood is the only event that can explain the paranormal features of earth's geological strata. For the reason that this is true, the geological episodes that took place "long ago" are not to be interpreted by the doctrine of uniformity but by what we see in the rocks. Our common sense observations, coupled with the revelation of God's word, should enable us to turn from the philosophy of uniformitarianism to the empirical facts verified in the catastrophic nature of the earth's stratums. Such layers could have been laid down only by the huge volumes of water which had their source in the biblical Flood.

CHAPTER THREE

THE GIRAFFE'S NECK
"THE LONG AND THE SHORT OF IT"
Natural Selection
(Survival of the fittest)

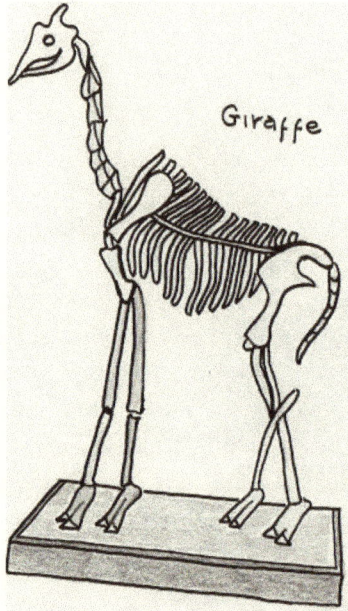

Giraffe

In the *National Geographic*, November 2004, P.7, David Quammen makes the following observation:

THE GIRAFFE INTRIQUED HIM [DARWIN] LESS FOR THE LENGTH OF ITS NECK THAN FOR THE SHAPE OF ITS TAIL, WHICH LOOKED TO HIM LIKE A "FLY-FLAPPER." FLY-SWATTING, HE NOTED, COULD HELP AN ANIMAL SURVIVE.

In the *Origin of Species*, P.144, Charles Darwin's commentary appears under the heading ORGANS OF LITTLE APPARENT IMPORTANCE, AS AFFECTED BY NATURAL SELECTION:-

"In the first place, we are much too ignorant in regard to the whole economy of any one organic being to say what slight modification would be important or not ... ...

The tail of the giraffe looks like an artificially constructed fly-flapper; and it seems at first incredible that this could have been adapted for its present purpose by successive slight modifications, each better and better fitted, for so trifling as object as to drive away flies; yet we should pause before being too positive even in this case."

Comments: - The reconstruction of the giraffe on page 7 of the *National Geographic* is spectacular. I appreciate the hard work that goes into the preparation and restoration of such skeletons. In visiting any museum, the first section I am likely to view is the area of Paleontology.

In thinking of the giraffe's skeleton, my thoughts are directed to the Creator and his plan to have all mammals with the same general skeleton design. The divine blueprint called for diversity and variation in the basic mammalian skeleton, while using the same main bones for each creature.

Let us return to David Quammen and his remarks concerning Charles Darwin. When he claims that Darwin was more interested in the giraffe's tail than in his neck, this had to be a tongue-in-cheek remark; a fly-swatting tail to help the giraffe survive? Darwin used anecdotes in his writing of *Origins*, but this had to be his crudest yarn on survival of the fittest.

Picture Darwin, if you will, leading St George Mivart (a distinguished zoologist of Darwin's day) to the site of a giraffe graveyard in Africa:

"What are these?" Mivart inquires.

Darwin responds, "These giraffes died because their tails did not meet the standard measurement and, being too short, failed to reach the flies hitching a ride on their rumps."

Mivart exclaims, "I know, don't tell me! Survival of the fittest! – Right? (These remarks indicate that it was only the long tailed giraffes that survived in support of Darwin's theory).

However, I think there was another reason for Quammen directing our thoughts to the giraffe's tail rather than to his neck. Could it be that

Darwin had made some rather unscientific remarks about the neck of the giraffe! For example:-

"So under nature with the nascent giraffe the individuals who were the highest browsers, and were able during dearths to reach even an inch or two above the others, will often have been preserved. ... the nascent giraffe, considering its probable habits of life; for those individuals which had some one part or several parts of their bodies rather more elongated than usual, would generally have survived. These will have intercrossed and left offspring, either inheriting the same bodily peculiarities, or with a tendency to vary again in the same manner; whilst the individuals, less favored in the same respects, will have been the most liable to perish."

Origins, P.161

It appears to me that Darwin, in his attempt to explain the way the neck was fashioned and passed on to offspring, went over to Lamark's theory of inherited habit. (Ibid. P.207)

Luther Sunderland comments with these relevant remarks on inherited habit:

"Charles Darwin devoted nearly a page in *The Origin* to a scenario that attempted to explain the origination of the neck of the giraffe. Although generally criticizing Lamarckism, in certain cases he accepted as an evolutionary mechanism the idea that offspring could inherit characteristics that the parents had acquired during their lifetime (Lamarckism). Using this idea, he speculated that the giraffe got its long neck by stretching higher and higher to reach leaves on trees as vegetation gradually dried up during a drought. There was more food on the highest branches and the least competition for them. The long neck was passed on to the offspring. Today, however, the neck of the giraffe is commonly given as the classic example of how wrong Lamarck was."

Darwin's Enigma

P.83

Did shorter-neck giraffes become extinct because their brothers developed long necks and were able to become the highest browsers? Did the offspring truly inherit long necks from the parents which acquired this characteristic in their lifetime? Is this truly science or the mere speculations of an over zealous naturalist? And, what would have prevented the short neck giraffes from taking a hike over the nearest ridge where food was more accessible? (There wasn't always a dearth). No need to die; plenty of food for everyone, only in another area. And what happened to the young giraffes? Did their mothers lift them up to the higher trees for sustenance or did they go to areas where there was more food?

Continuing with the subject of giraffes, Darwin said, "It seems to me almost certain that an ordinary hoofed quadruped might be converted into a giraffe."

Origins

P.161

Of course this process would take millions of years for the hoofed quadruped to step up on the phylogenitic tree and eventually become a giraffe. All this would happen, Darwin reasoned, through natural selection and survival of the fittest. But where in the fossil record are the intermediate forms that should be anticipated? Is there any evidence that the bones in giraffe necks were gradually becoming longer and longer and, today, shouldn't we witness also some living forms with medium sized necks?

The late Stephen Gould complained about his son's high school biology book begins the only chapter on evolution with the controversial issue of Lamarck and the inheritance of acquired characters. He said, "It then moves to Darwin and natural selection and follows this basic contrast with a picture of a giraffe and a disquisition of Lamarckian and Darwinian explanations for long necks."

Bully for Brontosaurus

P.165

Gould misdirects his readers by feigning a contrast between Darwin's "natural selection" and Lamark's theory of "inheritance of acquired characters" when both explanations for the giraffe's long neck were very much the same.

David Quammen, in my opinion, knew what he was doing by directing our minds to the tail of the giraffe and away from Darwin's unconvincing remarks concerning the neck of the giraffe; remarks that Quammen hoped we wouldn't discover by close investigation. This is simply another example of misleading, often employed by those who conform to Darwin's evolutionary doctrine and seek to cover up his unpredictability. Sunderland's striking statement will end this section on the Giraffe:

"Furthermore it is not possible for evolutionists to make up a plausible scenario for the origination of either the giraffe's long neck or its complicated blood pressure regulating system. This amazing feature generates extremely high pressure to pump the blood up to the 20-foot-high brain and then quickly reduces the pressure to prevent brain damage when the animal bends down to take a drink. After over a century of the most intensive exploration for fossils, the world's museums cannot display a single intermediate form that would connect the giraffe with any other creature."

Darwin's Enigma

P.84

CHAPTER FOUR

THE GIRAFFE'S NECK
"THE BIG 7"

Comments:-In the first situation, the giraffe is placed in the setting of natural selection and survival of the fittest. In the second example, the giraffe will have his cervical anatomy compared with some other animals. This branch of study is called morphology or comparative anatomy. Since I will discuss as a further example, the morphological comparison of the orangutan with man pictured on page 21 of the *National Geographic*, I will make my comments to the point.

Giraffes, mice, elephants, and porpoises have seven cervical vertebrae. Evolutionists love to talk about phylogenitic relationships and morphology in support of evolution when focusing upon the seven neck vertebrae. They seem to dismiss the problem of the lumbar and caudal spines which in different mammals contain a variation in the numbers of vertebrae. Nevertheless, the problem still remains. Variations within the spinal columns cannot serve as good morphological examples because they counteract the evolutionary theory of phylogenitic relationships – the very thing evolutionists set out to prove. This incongruity of reasoning not only demonstrates a contradiction of terms when it comes to anatomical morphology; it removes science from the investigation of facts and places it in the realm of speculative and worthless philosophy.

Nathan Fasten, in his *Introduction to General Zoology* claims:

"...morphology deals with the form and structure of organisms, and it affords some of the strongest evidence of organic evolution."

P.640

Creationists recognize that the basic significance of organisms having similar morphology <u>could mean</u> they are species related. However, Fasten's claim that such examples provide us with evidence for organic evolution, imply that similar morphology <u>must mean</u> a common ancestry.

Creationists reason that the Creator formed the different kinds of plants and animals as He spoke them into existence. In this phenomenon of nature, through divine intervention, God chose to make certain organisms similar in anatomical structure with no common ancestry proposed.

Actually, "morphe" in the word morphology means form, shape, and structure; "logy" means logic and in this instance is the "study of the science of forms and shapes." The evolutionist is mainly interested in how animals are similar and he sees these similarities as proof that they have a common ancestor. The creationist observes the same similarities but notes that the differences are so obvious that no species can be considered ancestral to any other. The fact of nature -"like comes from like" – is contradicted by the evolutionist's spin on comparative anatomy. The conclusion that similarities are evidence of a common ancestor is indirect and circumstantial. In the case of Darwin's organic evolution, scientists should expect to observe in nature the many varieties of organisms leading smoothly from one form to the next. But rather than research proving this expected result, the varieties of animals tend to cluster around a basic morphological pattern with large gaps between them. With the fossil record serving as the supreme and irrefutable example, organisms do not form a graded series but rather are divided up around patterns ("morphe").

What the creationist sees in nature serves as a strong argument for his position regarding basic kinds of animals. No matter how much variation can be demonstrated, there is always a marked difference between groups or kinds of animals. However, the position of the evolutionist, who disregards the many gaps in the fossil record, believes that organisms lead smoothly from one form to the next. He has the weaker argument since he does not believe, obviously, that animals form clusters around patterns ("morphe") and can, therefore, be distinguished from one another as "kinds". When his conclusions are based on what he does not see, then his research can only reap the fruits of pure philosophy which is far removed from the realm of empirical science.

Later, in this work, we will understand that even though men and apes may be very similar anatomically, they are not blood related. The creationist, within his philosophy, reasons that because man and beasts were formed from the dust of the ground and provided with similar foods for sustenance, certain similarities should be expected. Certainly, Darwin's philosophy which embraces the fact that man and beast are blood related and share a common ancestry is diametrically opposed to the creationist's philosophy which states that not only are beasts and man not ancestrally related but they are dramatically separated by man's moral, spiritual, and intellectual aspects as well.

Nature is not as simple as Darwin made it out to be. From evolution's standpoint, parallel structures in widely divergent forms may be homologous yet not indicative of a close relationship between organisms.

Darwin marveled:

" What can be more curious than that the hand of man, formed for grasping, that of a mole for digging, the leg of the horse, the paddle of the porpoise, and the wing of the bat, should all be constructed on the same pattern, and should include similar bones, in the same relative positions? ... ... Professor Flower .... Remarks in his conclusion: 'We may call this conformity to type, without getting much nearer to an explanation of the phenomenon…but is it not powerfully suggestive of true relationships, of inheritance from a common ancestor?'"

The Origin of Species

P. 334

When investigators of nature deem it logical to view the phenomenon of homology of the basic design of the forelimbs of terrestrial vertebrates as arising from the fact that they have an alleged common ancestry, is it not just as reasonable to believe that common patterns of organisms have their common origin from the one Architect or Designer?

It is not science which leads men on their investigations of similarities in organisms. The investigator, whether he is a creationist or evolutionist will have his evaluation colored by presuppositions. But as we will discover later, homology; the hierarchic pattern of nature; the aspects of comparative anatomy do not provide the circumstantial evidence that Darwin needed to prove his macroevolution claim. Neo-Darwinists claim that genetics is now a major part of their evolutionary theory. However, geneticists continue to affirm *limited change* of organisms – *microevolution* – take place in both nature and in the laboratory. This is rather stunning when it is known that neo-Darwinism teaches that major changes occur in organisms through the mechanics of *macroevolution*. Again, there is a contradiction between the realism of ordinary occurrences in nature and the fictitious beliefs that occur and rolls around inside the skulls of neo-Darwinists.

Again, we must note why the philosophy of the creationist is diametrically opposed to the philosophy of the evolutionist. The creationist holds the position that life's designs and patterns and its adaptive complexity is the result of purposeful activity on the part of God. Darwin and evolutionists take the stance that all life arose by chance mechanisms from a common ancestor cell. Basic design and aspects of comparative anatomy are the alleged proofs of this position. However, chance and design are antithetical concepts. It is difficult for the creationist to understand how the evolutionist can speak of design and chance in the same thought process.

Michael Denton makes this observation:

"By its very nature, evolution cannot be substantiated in the way that is usual in science by experiment and direct observation. Neither Darwin nor any subsequent biologist has ever witnessed the evolution of one new species as it actually occurs."

Evolution: A Theory in Crisis

P.55

Darwin's evolutionary theory, having to do with similar structures in organisms, was entirely speculative. In many pages to come, the reader's thoughts will be directed not only to this fact but to the idea that *the true laws of natural science are mostly held by the creationist.*

Michael Denton (*who is not a creationist*) in the last chapter of this same remarkable book, *A Theory in Crisis* makes this incredible statement:

"Put simply, no one has ever observed the interconnecting continuum of functional forms linking all known past and present species of life. The concept of the continuity of nature has existed in the mind of man, never in the facts of nature. In a very real sense, therefore, advocacy of the doctrine of continuity has always necessitated a retreat from pure empiricism, and contrary to what is widely assumed by evolutionary biologists today, *it has always been the anti-evolutionists, not the evolutionists, in the scientific community who have stuck rigidly to the facts and adhered to a more strictly empirical approach.*"

Pp.354-355 [Emphasis mine]

Creationists stick to what they are able to observe in nature. Evolutionists stick, by faith, to their philosophy of evolution - what cannot be observed in nature. No wonder that nearly all leading evolutionists have stopped debating these issues publicly on the campuses of major colleges! Their peers have asked them to refrain from such debates because it places evolution in an unfavorable light within the eyes of the students.

CHAPTER FIVE

THE ROCKDOVE PIGEON
AND THE ALBINO-
"THERE ARE LIMITS IN THE GAME OF VARIETY"

Albinism

Flying Fish

RBP
34

In the *National Geographic*, Pp. 10-11, there is a photograph of a rock dove with the flesh boiled off. David Quammen claims that Darwin was a "HANDS-ON-EXPERIMENTALIST."

It is extremely interesting to take note of the variations which arose from the ancestral form of pigeons such as the English pouter, the nun, the fancy pigeon, etc. Nevertheless, Darwin was very much aware that in his experience as a breeder, he started with pigeons and ended with pigeons. This material is repeatedly used by evolutionists as proof that evolution of new kinds is going on and arises in our day.   Contrariwise, no evolution of new kinds is occurring. The best explanation is still, each "after his kind." Similarities or variations in a particular kind of pigeon do not even suggest evolution of one kind into another. Pigeons descended from pigeon-like ancestors and they, in turn, will give rise to pigeon-like offspring. Not a thing more than this can possibly be expected from pigeons.

David Quammen, on page 12 of the *National Geographic*, makes the following comment:

MUSTERING FACTS FROM MANY DIVERSE REALMS, DARWIN SAW THE IMPLICATIONS OF ALBINISM AS AN INHERITED TRAIT.

Comments:-Albinism can affect all human races. Albinos have pink or light-blue irises, fair hair, and abnormally pale skin. Darwin saw albinism as an inherited trait and so it is. Albinism is an inherited disorder in which tyrosine, one of the enzymes required for the formation of the pigment melanin, is absent. I suppose the implications seen by Darwin in the inherited trait, were real and imaginative. He observed the real situation, an individual who is different from his parents; in imagination, Darwin saw this mutational defect as the major driving force behind evolution. A *mutant* defined, is an individual or organism differing from the parental strain as a result of mutation. A *mutation* is any heritable alteration of the genes or chromosomes of an organism.

The following are my notes written at the bottom of pages 12-13 of the *National Geographic*. The notes appear in my personal copy:

"Yes, albinism is an inherited trait in which a person cannot produce melanin, the coloring pigment in our skin. The implication, apparently, was the mutational-caused defect. But this is a minor mutation within the race of mankind causing an individual to differ from the parental strain. An alteration of the genes or chromosomes will produce a mistake leading to variation but not unlimited variation, as Darwin believed."

Like the evolutionist, creationists believe in variety. Both groups look to gene change for diversity among species. The evolutionist believes that mutations are the greatest possibility for evolutionary change from one species into another; the creationist believes that gene changes are the main source of the multiplicity of variation seen within plants or animals.

So, both groups believe in variation. What, then, is the difference between the two groups? The evolutionist believes, through modification or mutational change, one type of animal will eventually develop into another type, completely diverse or unlike the first. However, the creationist believes that although the evidence for gene variation within plants or animals can be extensive, it is never enough to bridge the gap between the old kind and the new kind. In other words, there is no evidence that mutational change is able to produce a completely new plant or animal from the old.

The crucial law in nature is each kind is set off from every other kind by some "residual part" which no amount of gene change can erase. The greatest changes that the creationist sees (but many evolutionists fail to acknowledge) are the mere production of additional modern "species" within groups already set off in nature. There will be more about mutational change under the title "The Genetic Revolution" and mentioned on pages 32-33 in the *National Geographic*. In this book, the same subject (mutational change) will be covered in chapter 16 – "The Fruit Fly---Once a Drosophila Always a Drosophila."

CHAPTER SIX

THE GLIDING FISH WHO DIDN'T
ATTEND FLIGHT SCHOOL

Flying Fish:

DARWIN OBSERVED THE WINGS OF A FLYING FISH.
ALTHOUGH THE FISH'S WINGS ARE RUDIMENTARY
COMPARED TO A BIRD'S, HE REALIZED THAT THEY
DERIVE FROM THE SAME EVOLUTIONARY PROCESS:
THEY ENABLE THE FISH TO SOAR TO ESCAPE PREDATORS.
National Geographic, P. 12

Comments:- Before continuing with the flying fish story, I have to say that David Quammen is either ill-informed or which is more apt to be accurate, he doesn't care what he says so long as he is able to push his evolutionary theory. We will discover that flying fish did not have wings nor were they, in fact, rudimentary. There are about 40 species of the flying fish. They make up the family Exocoetidae and are up to 18 inches long. The flying fish swims just below the surface in warm oceanic waters. If disturbed by predators, they launch themselves from the water. Quammen claims they do this by rudimentary wings. He is wrong on both counts. First of all in claiming that pectoral fins are rudimentary and secondly by calling the large fins, wings.

The flying fish launches itself from the water by rapidly beating its tail and glides through the air by using its large pectoral fins. The flying fish does not have wings but rather fins which it uses for gliding. The fins are not rudimentary in the sense of being underdeveloped wings. Rather, the fish has a double use for its large, totally, developed fins; they not only assist it in swimming below the surface of the oceanic waters; they also help the fish to glide through the air when there is a disturbance in the water.

Meanwhile, here are a few interjectory thoughts on Charles Darwin and his positive qualities. From the time I was a teen-ager until now, I have read *The Origin of Species* and *The Descent of Man* five times. Firstly in order to be certain that I understood Darwin's opinions about evolution and secondly, learn the particulars about zoology. There are noteworthy specifics about animals and their habits to be gleaned from the writings of Charles Darwin. His character assets can be cited in the following manner: his writings of books on geological subjects - coral reefs, volcanic islands, and the geology of South America; his keeping of notebooks in a systematic way; his accomplishments, in spite of his being an invalid for much of his life; his obsessive thoroughness; his trekking for weeks across mountains and deserts in looking for fossil bones; his seeking first-hand information from breeders, gardeners, and zookeepers (Gould); finally, his calling attention to variation – a trait actually existing among living things. Darwin was not the first to call attention to this attribute subsisting in organisms. This was an empirical detail of nature which could be deduced from simple observation. It was a fact known by all scientists but it should be admitted, he helped to remove focus from the unscientific position of the scholastic instructors who taught fixed species.

However, there are aspects of his character which make me uncomfortable: I recently learned of Darwin's passion for hunting and killing and his killing bordered on sadism; his collecting living specimens and not taking time to kill them painlessly; his intellectual prowlness in holding up natural selection as a mechanism of evolution when such a theory was based not on facts he had collected but on a faulty exposition of those facts; his complete disregard of the personal copy of Gregor Mendel's findings; his disregard for the book of Genesis and in discovering the truth about fixed species, deduced that by refuting the scholars who taught Genesis incorrectly, he had in turn proved the Bible to be inaccurate; his lightning-quick power of deduction which got him into trouble and especially when he published such deductions. For example, the view he had of flying fish: "...and seeing that we have flying birds and mammals, flying insects of the most diversified types, and formerly had flying reptiles, it is conceivable that flying-fish, which now glide through the air, slightly rising and turning by the aid of their fluttering fins, might have modified into perfectly winged animals."

The Origin of Species

P.130

For Darwin to assume that the flying fish was a possible candidate for future flight training what is more than likely to receive its 'wings' award, is not only ludicrous but just plain asinine. This is where the belief in the power of modification oversteps the point of credibility. Evolution cannot accomplish everything simply because it waves an imaginary magic wand, no matter how many million of years are the perceived time. What kind of "perfectly winged animal" might Darwin suppose this flying fish would turn into? Did Darwin surmise its turning into a flying insect such as the largest dragonfly with a wingspan of 24 inches; a bat with its marvelous sonar system; a bird, with its intricate system of characteristics designed for the air; a flying reptile, such as the extinct Pteranodon or Rhamphorhyncus? Was Darwin thinking of another animal to be added to the four we known about. If he was, we can't help but wonder what winged animal Darwin might have imagined!

For fear of wasting further time on Darwin's absurd remark on flying-fish, one more important thing should be mentioned for emphasis and in a number of different ways: - There were no modifications (resulting from genetic changes) in any animal ancestors within the fossil record, leading up to any of these four flying animals; no transitional fossil forms leading up to any of the flying species; no transitional precursors to usher in flying insects, bats, birds, nor flying reptiles. With nature's demonstrations, showing the absence of modification between kinds of animals in the fossil record of Darwin's day, how could he possibly have believed in the fantastic notion of a flying fish being modified into some kind of "perfectly winged" animal? Again, Darwin's evolutionary philosophy was simply a matter of faith. It can be said of this evolutionist – his "Faith was the substance of things hoped for, *the evidence of things never to be seen.*"

CHAPTER SEVEN

MAMMALS, SNAKES, AND BEETLES
AND THEIR
"NON-FUNCTIONING" ORGANS

In the *National Geographic*, November 2004, David Quammen, at the top of pages 12-13, has the following questions in bold print:

WHY DO MALE MAMMALS [including human males] HAVE NIPPLES?

WHY DO SOME SNAKES CARRY THE RUDIMENTS OF TINY LEGS?

WHY DO CERTAIN SPECIES OF FLIGHTLESS BEETLES HAVE WINGS THAT NEVER OPEN?

Comments:-Quite a few vestigial structures have been selected for the reader's consideration, particularly those structures mentioned above.

Richard M. Ritland defines vestigial structures this way:

"Vestigial organs or parts of the body are supposed to be the vestiges or remnants of organs which have in the course of evolution lost their primary function and been reduced usually to a fraction of their original size."

A Search for the Meaning in Nature

Richard M. Ritland

P.290

Such characteristics by evolutionists are considered to be "functionless remnants" or "empty gestures to the past." Evolutionists cite these "empty gestures" as testimonials against the perfect work of creation and hoping to substantiate that nature is not the result of a Divine Architect following His blue-print but conversely confirming that life erected itself through the blind forces of chance labeled spontaneous generation, variation without limitation, natural selection, and progressive evolution.

Evolutionary scientists assert that about 180 human bodily organs are considered to be no more than useless vestiges. They make this declaration firstly in an attempt to prove that there could not be a Creator due to the imperfections in nature and secondly, to prove that vestiges are supposed to be remnants or left over parts from the course of man's evolutionary ascent from his animal ancestors.

Over the years many valid, scientific discoveries have been made finding uses for these so-called vestigial organs and the list has been considerably shredded. However, scientists still cite the old tabulation to pad their supporting argument for the doctrine of evolution. Let us proceed by taking up the issue of vestigial structures and how they are employed by evolutionists as a confirming factor for their theory. Some of the "rudiments" in man and animals will be considered.

Man: The coccyx is cited as a functionless remnant of the embryonic "tail" in man. Nevertheless, this structure helps support the pelvic cavity; furnishes a surface for the attachment of the large gluteus muscle which rotates the thigh; protects the terminal portion of the spinal cord; the skin around the coccyx is supplied by the coccygeal nerve which finds an outlet in the coccyx; is difficult for a person to sit down, once the coccyx has been surgically removed; aids in defecation; is necessary for good posture and support.

WHY DO MALE MAMMALS [including human males] HAVE NIPPLES?

Male nipples (in humans):- The only thing that guides us in our response to this issue is the observation of the facts in nature - the observation which, of course, is governed by our own personal prejudices. The evolutionist, who tells us that we must approach nature with a blank mind in figuring out its history, is not exactly straightforward. Every person, through his experiences in life, is conditioned to interpret nature's facts a certain way. Take for example the creationist and the evolutionist. As a creationist, a person might answer the question why do males have nipples in the following way: Male nipples are a part of the design economy initiated by the Creator. When He set up the plan for early embryonic development, both male and female embryos start out by producing those features which they have in common. The evolutionist, however, reasons differently in his approach to natural history. Darwin ridiculed the creationist by writing, "...rudimentary organs are generally said to have been created 'for the sake of symmetry', or in order 'to complete the scheme of nature'; but this seems to me no explanation, merely a restatement of the fact."

(Darwin's quote is in the *Structure of Evolutionary Theory* by Stephen Jay Gould, P.112)

On the one hand, the creationist looks to the perfection in nature as pointing to the Great Master Designer. On the other hand, the evolutionist looks for examples from the oddities and imperfection in the realm of nature. He claims that rudimentary and homologous structures point to nature's scheme of modification that has nothing to do with harmony and design.

The argument is set forth by creationists that embryos start out producing features that are common to both male and female. But what do males and females have in common when it comes to mammae? The undisputable answer to this question remains uncertain but nevertheless; both male and female begin their embryological development in this fashion. This is a fact of science and the idea about the design economy in nature may be valid.

Darwin suggested that long after the progenitors of the whole mammalian class had ceased to be androgynous; both sexes yielded milk, and thus nourished their young. This thought shows an interesting but extreme viewpoint. To carry his thinking further, Darwin even mentions that the mammary glands and nipples in male mammals can hardly be called rudimentary since they are merely not fully developed and not functionally active. He writes, "They often secrete a few drops of milk at birth and at puberty."

The Origin of Species

Darwin, P. 521

For my own thinking, I am inclined to believe this example that Darwin pointed out, is not the usual order of things in nature. For him to make a case out of mammals at one point being sexless is extreme to say the least. As a creationist, how can I answer why do males have nipples? I would first point out that nature makes it clear to me that there are no so-called "rudiments" pointing back in time to previous evolutionary ancestors. This is evidenced by the fact that true rudiments are degenerative. Subtractions in DNA cannot produce new characteristics in organisms. And furthermore, nature does not relate to us the full revelation of why males have nipples but this I do know - the list of so-called "rudimentary organs" is growing shorter over the years.

With knowledge coming to the front, man is finding "rudimentary organs" to have definite functions. Evolutionists should be cautious in stating that a certain organ has no functions whatsoever, especially when many such organs prove otherwise. Creationists must also be cautious when seeking to establish a *function for every organ* since there are a number of organs that *are truly vestigial* in nature. For example, blind cave fish. In general, blind cave fish develop what appear to be normal eyes in their embryonic history but later lose them through atrophy. The "degeneration" of the eye is possibly the result of a single mutational change and the resulting structure is definitely a vestigial one. It is not surprising to me that along with the sin-problem, more structures of the original creation are not actually degenerated and vestigial.

I can not come up with an authoritative answer to the question why do males have nipples. But I would posit a question for the evolutionists to answer-"Of what real value are such structures to your theory?" Most evolutionists see the development of life as the more complex from the simple. Truly, vestigial structures are subtractions, not additions. Some creationists have coined the term "devolution" to describe deterioration or degeneration since the fall of man. I believe that is a valid way of describing the loss of features by natural process; of describing the loss of DNA information; of expressing the impossibility of the alternative evolutionary theory that, in some mystical way, induces scientists to believe a subtraction in DNA can add up to new characteristics within the species and provide evidence for microbe-to-man evolution.

In conclusion, on the question of male nipples, it may be that the answer by most creationists that nipples are the result of design economy initiated by the Creator will have to suffice. The future might hold still another answer as to the use of male nipples. Meantime, creationists have to content themselves with the knowledge that the question of vestigial structures offers no supporting evidence for the theory of evolution. It is of a truth, vestigial structures are indicative of a gene loss and not an addition. The evolutionist can not subtract genes and, at the same time, expect to add new characteristics to animals in their alleged progressive climb to the top of the evolutionary pinnacle. And now, returning to further organs in the human anatomy ... ...

The human appendix was once cited as a totally useless organ. Nevertheless, the appendix contains lymphatic tissue and helps control bacteria entering the intestines; that is, the appendix produces white cells and performs a significant role in the production of antibodies.

The tonsil once commonly cited as a vestige, like the appendix, is largely composed of lymphatic tissue and produces white cells in warding off infection. Years ago I had my tonsils removed because of certain evolutionary views which, due to ignorance, discounted the importance of tonsils and their usefulness in providing antibodies.

The group of small muscles which move the external ear and the scalp above the ear are claimed to be vestigial. In animals, the muscles are larger to meet the requirements of motion of the ears that aids in the animals acute sense of hearing. According to the evolutionist, ear muscles in man are cited as vestigial because he reasons; they are no longer needed as they once were in his animal ancestry. However, the muscles, although limited, do contribute to facial expressions and movement of the scalp. Any bodily system or structure is adapted to meet the particular requirements of an organism in his particular biological setting. In this case, man has limited ear muscles simply because his needs are limited. (Page 293 in Richard M. Ritland's book – *A Search for Meaning in Nature*)

"Pseudogenes" is listed by Douglas Futuyama as an example which confirms the creative effectiveness of natural selection. He takes his clue from Ken Miller who claims that the cluster of five genes that produce hemoglobin in humans, speaks against the theory of intelligent design or a Creator. Miller claims that there are five genes to which a fifth "pseudo gene" was added. This fifth gene was a mistake of which evolution took advantage. His argument, in summary, runs this way: {there are five genes which produce different forms of hemoglobin in humans. To this cluster of genes, one gene has been added that is nearly identical to the others. Oddly, however, this gene plays no role in producing hemoglobin. Biologists call such a region a "pseudo gene." The pseudo gene lacks proper signals to inform the rest of the cell's machinery to make a protein from it.}

Miller concludes that if a Creator designed this, then He must have made some serious errors in wasting millions of bases of DNA on a blueprint full of junk and scribbles. He then goes on to explain that evolution can easily explain them as failed experiments in a random process of gene duplication.

These wasted "evolutionary remnants" are simply a molecular version of the argument for vestigial organs. Michael J. Behe, professor of biochemistry at Lehigh University in his book *Darwin's Black Box*, Pp. 226-228, meets Miller's claim. Behe shows this argument is inconsistent for a number of reasons: because we have not yet discovered a use for a structure does not mean that no use exists; the tonsils, once considered to be useless organs, have an important function in immunity - this point also applies on the molecular scale; pseudo genes, although they are not used to make proteins, may be used for other things that we don't know about. The point that is made here is Miller's assertion rests on assumptions only. Why Miller's argument fails to persuade is that even if pseudo genes have no function, *evolution has "explained" nothing about how pseudo genes arose.* Miller cannot inform us as to how these functions might have arisen in a Darwinian step-by-step process.

Behe offers this illustration for design-"If I insert a letter into a photocopier for instance, and it makes a dozen good copies and one copy that has a couple of large smears on it, I would be wrong to use the smeared copy as evidence that the photocopier arose by chance." (Pp.226-228)

Behe argues further and then closes his remarks by stating, "When it is considered by itself-away from logically unrelated ideas-the theory of intelligent design is seen to be quite robust, easily answering the argument from imperfection." (P.228)

Any evolutionist, after reading "Darwin's Black Box," who comes away from this reading and believes that the science of biochemistry still supports the theory of "no-design" and "no - Intelligence," hasn't read the book with an open mind.

Animals:

WHY DO SOME SNAKES CARRY THE RUDIMENTS OF TINY LEGS?

For example consider the Boa and Python family (Boidae). Reading from *The Audubon Society Field Guide to Reptiles*, P.586, there is this descriptive account of the hind limbs... ... "Vestiges of hind limbs are present as 'spurs' usually visible on either side of the vent." The Python, Boa, and most other members of the Boidae family of snakes carry small bones. They are frequently listed as appears in the above description, vestigial structures. In some accounts, it is more specific that these structures have no probable function.

However, it comes to light that these claw-like spurs move and are used in courtship and reproductive functions. *Thus, they most definitely have functions.*

Rhea, as shown on P.22 of the *National Geographic*, was a large flightless bird. John Gould, Darwin's ornithological consultant, on March 14, 1837, presented a paper at the Zoological Society on this large bird collected by

Darwin in southern Patagonia. Gould, in considering the bird, thought it represented a new species which he renamed Rhea darwinii. Darwin's interest in the flightless bird was obviously aroused. The evolutionist points to this Rhea, the Ostrich, and other flightless birds and claims the wings are vestigial.

The Answers Book offers an excellent array of wing functions for flightless birds. For examples: balance while running; cooling in hot weather; protection of the rib-cage in falls; mating rituals; scaring predators; sheltering of chicks. Also, if the wings are useless then why are functional muscles allowing these birds to move their wings? (*The Revised and Expanded Answers Book*, P.124)

Let us consider one more question on animal organs: "WHY DO CERTAIN SPECIES OF FLIGHTLESS BEETLES HAVE WINGS THAT NEVER OPEN?"

Darwin writes about flightless beetles:

"Rudimentary organs plainly declare their origin and meaning in various ways. There are beetles belonging to closely allied species, or even to the same identical species, which have either full sized and perfect wings, or mere rudiments of membrane, which not rarely lie under wing-covers firmly soldered together; and in these cases it is impossible to doubt, that the rudiments represent wings."

The Origin of Species

P.346

Darwin had a passion for natural history; he kept a beetle collection from the time of his childhood; he wrote *The Origin of Species* inspired by this same childhood insect collection. Nevertheless, issue can be taken with him on the subject of beetle wings as rudimentary organs. The beetle is an insect belonging to the largest order of the animal kingdom (Coleoptera; about 278,000 species). According to evolutionists, insects were the first creatures to fly and according to most books on entomology, there is always an apologetic statement such as "but the origin of wings is not clearly understood." This is not surprising since the origin of anything having to do with animals is impossible to explain using the evolutionary model. This is why Darwin's book on *The Origin of Species* (374 pages) never once proved origins; an issue that evolutionists fail to prove in their books.

Repeating the candid remark of evolutionists, "But the origin of wings is not clearly understood" might be added "especially beetle's wings." The wings of the beetle are an engineering marvel. The cockchafer beetle can be used as an example: before it takes off, the beetle must warm up. He opens and shuts his wings to make certain that they are in good working condition; he unfastens the wings by separating the hardened wing cases from the more delicate hind wings; he spreads his antennae to check for air currents; his feet claws grip the plant firmly before taking off; his thin membranous hind wings provide the propelling force; as the speed increases, the beetle's front wings provide aerodynamic lift (The above description of the cockchafer beetle is in "Eyewitness Books Insects," Pp. 12-13, written by Laurence Mound). In thinking of the complexity of beetle wings, no wonder the origin of such a system is left unexplained.

With the importance of flight enabling the beetle to escape from predators or to fly to new areas in search for better food, why are there flightless beetles? Since most beetles fly, we can well assume this to be a genetic deterioration influenced by some mutational change. Mutations are often lethal and *may* cause the death of any mutant beetle. The beetle mutant can be the result of wing deterioration. Nevertheless, *the genetic change will not necessarily mean the mutant's annihilation.* In the case of the flightless beetle, loss of wings by mutation might appear to be an impossible handicap, yet wingless mutants of winged forms of beetles "find life possible on windy oceanic islands while their *winged* brothers and sisters are blown out to sea" (Marsh). "Happy days are here again" for the beetle mutants.

Considerable time has been spent in answering David Quammen's questions on vestigial structures but we need one minute for a brief closing.

The Vestigial Structures Argument:

There are three main arguments against the "supporting" evolutionary theory of "vestigial structures." 1) All but few "vestigial structures," cited by evolutionists over the years, have been debunked - that is, they have been exposed as having useful functions; 2) the few "vestigial structures" that remain on the evolutionary list, cannot possibly be proved conclusively as having no function at all; 3) finally those structures which are "truly vestigial" are the result of mutational deterioration.

From the point of view cited in this chapter, it should be understood that vestigial organs fit better into the pattern of conservative change and degeneration than of progressive change and improvement.

CHAPTER EIGHT

THE MOTH AND THE LILY-
FRIENDS FROM MADAGASCAR

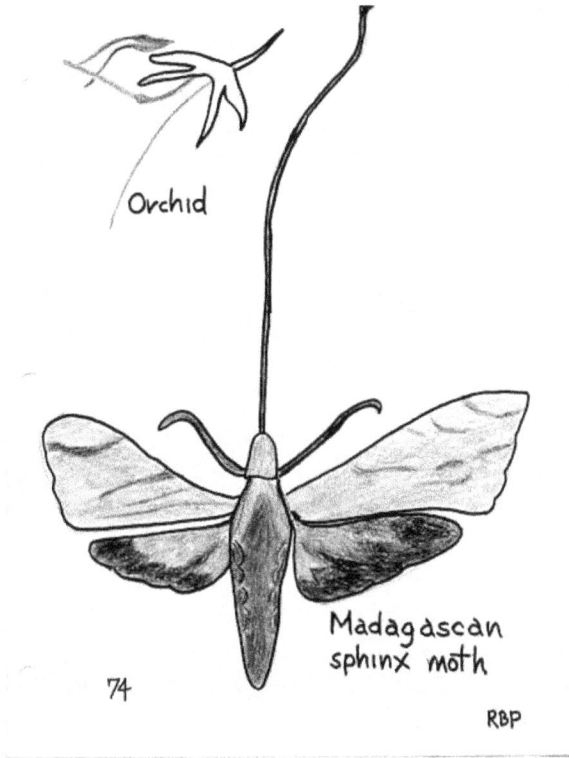

Orchid

Madagascan
sphinx moth

74

RBP

In the *National Geographic*, November 2004, Pp. 14-15, David Quammen URGES US TO SEE LIKE DARWIN. Darwin observed the Madagascan orchid and the sphinx moth in mutual adaptation which is called "co-evolution." However, does this mutual adaptation have anything to do with evolution?

Comments: - I am glad that I do not see things in nature the same way that Darwin did. On the contrary, if I were living in his day, he would have considered me too blind in my understanding of nature. He wrote:

"Nature may be said to have taken pains to reveal her scheme of modification, by means of rudimentary organs, of embryological and homologous structures, but we are *too blind* to understand her meaning." [Emphasis mine]

P.367

This Darwinian statement contrasts with that of the Apostle Paul's statement in his letter to the church at Rome:

"The wrath of God is being revealed from heaven against all the godlessness and wickedness of men who suppress the truth by their wickedness, since what may be known about God *is plain* to them. For since the creation of the world God's invisible qualities-his eternal power and divine nature-have been *clearly seen*, being understood from what has been made, so that men are without excuse."

Romans 1:18-20 NIV.

Darwin *saw* natural selection throughout all nature as the powerful force for evolution. Paul *saw* the invisible qualities of God behind all nature, as the powerful force for creation. Darwin wanted us to *see* that the natural world arose by accident rather than by design. Paul wanted us to *see* that the world we live in did not come about by mathematical chance or accident but was the outcome of Divinity who spoke this world into existence with all its designed forms. It is by virtue of the things that are made and fashioned; we understand God as being the true God; opposed to heathen gods which had no power to create man and in fact were created by man. Paul said that men are without excuse for *seeing* the wrong things in nature. If he were writing today, some of the wrong things Paul would have noted are: Darwin's modification, rudimentary organs, and his evolutionary views on embryological and homologous structures. Paul was a most ardent creationist.

The apostle in addressing the church at Colosse wrote of Deity:

"For by him all things were created: ... all things were created by him and for him. He is before all things, and in him all things hold together" [God holds the atoms together. Otherwise things would separate]

Colossians 1:16-17.

Paul also wrote to the Hebrews:

"God ... ... in these last days he has spoken to us by his Son, whom he appointed heir of all things, and through whom he made the universe. The Son is the radiance of God's glory and the exact representation of his being, sustaining all things by his power and word ..."

Hebrews 1:1-3.

Darwin, in 1862, wrote a book about orchids being fertilized by insects. But he and the late Stephen Jay Gould uses this interdependency as a compendium for teaching optimal design, induced by evolution, and mimicking "the postulated action of an omnipotent Creator" (Gould remarks are in *The Panda's Thumb*, P.20). The charge was made that God would never have created such a flower with continued self-fertilization and short-term survival. It took evolution to create the necessary "contrivances" to attract insects and insure long-term survival. And so, two evolutionists downplayed the Creator and sat in judgment upon Him who lives forever.

ORCHIDS, WONDROUSLY ADAPTED FOR CONTROLLING THEIR POLLINATION BY INSECTS, INTRIGUED DARWIN.THE PARTS OF THEIR STRANGELY MODIFIED FLOWERS, HE SAW, CORRESPOND TO THE FLOWER PARTS ON SIMPLER PLANTS, SUGGESTING EVOLUTIONARY CHANGE.

National Geographic, P.14

Comments:-If the evolutionary premise in the final portion of the above statement is true, that complex forms of plants evolve from simpler ancestors, then why is there no evidence? Progressive evolution is not seen in real life or in the fossil record.

Daniel Axelrod, in *The Evolution of Flowering Plants*, makes this statement:

"The fossil record does not present a picture of gradual evolution of flowering plants from simple ancestral families."

(Axelrod's statement is quoted in *A Search for Meaning in Nature*, Richard M. Ritland, and P.142)

Take note that the above statement of a modern evolutionist disagrees with the observations of Darwin and Quammen!

ONE SPECIES THAT CAUGHT HIS EYE WAS THE MADAGASCAR ORCHID... (And the)...MADAGASCAN SPHINX MOTH... ...SUCH MUTUAL ADAPTATION –THE MOTH TO THE FLOWER, THE FLOWER TO THE MOTH-IS CALLED COEVOLUTION.

Comments: - I like to think of co-evolution as replaced by cooperation. I admire God's plan of active cooperation between organisms. In this fallen world and even with Darwin's claim that there is a day to day struggle for existence between plants and animals, there are numerous examples of tranquility and harmony in nature. The love of God still overshadows Darwin's evolutionary credo which embraces the "tooth-and-claw" philosophy:

"Despite Darwin's statement to the contrary, active cooperation is the rule, not the exception in nature."

Grand Canyon Monument to Catastrophe edited by Steven A. Austin

Pp.156-159

Additional comments on *seeing* like Darwin: - Darwin believed that natural selection, over many millions of years, could change species in such a way as to cause them to diverge from their common ancestors and to formulate entire new species. On page 56 of *The Origin of Species* he made this statement:

"We see nothing of these slow changes in progress, until the hand of time has marked the lapse of ages, and then so imperfect is our view into long-post geological ages, that we see only the forms of life are different from what they formerly were."

But Darwin was not the only scientist who studied natural selection. A number of writers saw natural selection in a different light. Harold Clark says:

"Here, by the way, is where natural selection plays its part, not in producing new species, but in maintaining something already formed and adjusting it to new conditions. A number of writers mentioned the concept of natural selection before Darwin took it up, but they saw it as something that preserved existing species, not a process that produced new ones"

New Creationism

P.29

Darwin mentions the scientists of his day and resisted their views:

"Several writers have misapprehended or objected to the term Natural Selection. Some have even imagined that natural selection induces variability, whereas it implies on the preservation of such variations as arise and are beneficial to the being under its conditions of life."

Origin of Species

P. 64

Darwin observed "natural selection" as did other scientists, but he saw it with his vision obscured by the darkness of ignorance. He chose this darkness when he failed to respond to the paper sent to him by Gregor Mendel on variation. Darwin could have learned something vitally important from the humble monk. Mendel demonstrated through experimentation that offspring may demonstrate characters by either parent but that it cannot develop any characters which were not manifest or latent in the ancestry.

Darwin could have been well acquainted with the idea that there is no such thing as the transmission of acquired characteristics; could have known that his theory of gradual changes in species, induced by natural selection until a wholly new form is produced, was vanished through Mendel's experiments; could have known, with a little coaching from Mendel's papers, that true variations were confined to narrow limits; could have known small variations cannot be accumulated into large differences equal in value to a new species.

Darwin still maintained that natural selection had power to produce new variants. Since the dawn of scientific observation, scientists have not once observed a new species originate by natural methods in nature or in the laboratory by artificial method. Michael Denton will end this section with a crucial negative against natural selection:

"It was not only his general theory that was almost entirely lacking in any direct empirical support, but his special theory was also largely dependent on circumstantial evidence. A striking witness to this is the fact that nowhere was Darwin able to point to one bona fide case of natural selection having actually generated evolutionary change in nature, let alone having been responsible for the creation of a new species."

Evolution: a Theory in Crisis

P.62

Readers! Do not take Quammen's advice when he urges you to *see* like Darwin! Darwin never *saw* the creation of a new species arising from the mechanics (?) of natural selection and neither has anybody else – ever!

CHAPTER NINE

PIGEONS AND BULLDOGS-WHAT
THEY HAVE IN COMMON WITH OTHER ANIMALS

In the *National Geographic*, November 2004, David Quammen refers to the bulldog and other examples of variation as proofs for evolution (Pp. 16-17). He makes the following comments:

THE BULLDOG, SHAPED BY MANY GENERATIONS OF DOG BREEDERS...DIFFERS MUCH FROM ITS WOLFISH PROGENITORS. IF DOMESTIC BREEDING COULD YIELD SUCH CHANGE, DARWIN REALIZED, NATURAL SELECTON OVER MANY MILLIONS OF YEARS COULD DO MORE.

Comments: - The action of natural selection is of great importance in the separation of variety. However, natural selection takes variants from the accumulated stock of change but has no power in itself to produce new variants. It cannot induce mutations nor does it bring about favorable gene combinations. Natural selection eliminates the less fit and prospers the more fit but it does not create new basic types. Natural selection can accomplish the survival of the fittest but not its arrival. (*Life, Man, and Time* by Frank Lewis Marsh, Pp.170-171)

Even when variety does take place through mutational change or a reshuffling of existing genes, it is not natural selection which induces the change no matter how many millions of years were in accordance with fact. Mutations are mistakes made during the transfer of information from the genes of one generation to the next. Therefore, mutations can alter structures but never create new ones. Likewise, recombination does not add anything new to the gene pool; it just simply makes use of the genes already in the gene pool itself.

HE [DARWIN] ARGUED THAT WILD SPECIES DIVERGE FROM COMMON ANCESTORS JUST AS DOMESTIC VARIETIES DO. USING HIS OWN BACKYARD AVIARY, AS WELL AS INFORMATION FROM OTHER BREEDERS, HE ANALYZED DIFFERENCES AMONG FANCY PIGEONS SUCH AS THE ENGLISH POUTER, THE SCANDAROON, AND THE NUN.HE ALSO STUDIED CATS, HORSES, PIGS, RABBITS, DUCKS, AND OTHER LIVE STOCK.

Comments:-And what did Darwin learn from studying all these varieties?
. . .

A mother cat always gives birth to kittens (Cats remain cats, although changing from tabbies to lions); that a colt always has a mare as its mother; that mother pigs always give birth to piglets, etc. Certainly, in Darwin's hands-on experiments, he must have observed that they resulted in varieties of species only. But he came to the mistaken, imaginatory impression that varieties are without limitation. The theory of evolution resulted from the thinking of a naturalist thinking in an unnatural way. Nature shares no empirical evidence that variety, incurring in species, is unlimited. The theory of evolution resulted from Darwin's unscientific imagination superimposing itself over what is the scientific truth in nature. Evolution is the end result of a man who built his observations on the shifting sands of fantasy rather than upon the solid rock foundation of cautious investigation. Frank Lewis Marsh speaks eloquently with reference to Darwin's weakness:

"His weakness lay in his inability to observe that variation did not occur in unlimited fashion. If he could have seen that the theory of creationism merely stated that the separate kinds could but remain separate, and that, with all the variation that occurs, nothing more is erected than additional races and varieties within the created kinds, he would have saved the scientific world its tragic pilgrimage into the delusion of unlimited variation leading to evolution of kinds, a delusion which still blows its confusing breath upon the scientific world of our day ... ...

"If he had held fast to the theory of creationism while recognizing these forces of change, he would have spared us the pathetic sight of our day as scientists spend their lifeblood searching vainly for missing links and for forces which will erect new kinds."

Evolution, Creation, and Science

Pp.215-216

HE EXAMINED AND MEASURED SPECIMENS, ALIVE AND DEAD. TO A FRIEND, HE [DARWIN] WROTE, "I HAVE PUPPIES OF BULL-DOGS AND GREYHOUND IN SALT."

Comments:-How earth shattering! -When Darwin revealed that the puppies he kept in salt came from Bull-dogs and a Greyhound ... dog-kind, if you please! Again, each organism still reproduces "after his kind." Dogs remain dogs, although genetically changing from great Danes to poodles.

The 1940 story of Goldschmidt, a biologist, sharply criticizing the Darwinian theory of natural selection in a book named the "Material Basis of Evolution" and, one night, lecturing on the same topic, was asked by a number of students the next morning to go over to the Museum of Vertebrate Zoology.

In an effort to prove evolution they laid out series of different kinds of animals. They pointed to variation and with assuredness; asked Goldschmidt if that did not imply evolution. In answer to their question Goldschmidt responded that he indeed observed variation but noted that rats were still rats, the rabbits were still rabbits, and the foxes were still foxes. He saw no evidence of one kind turning into another (*New Creationism* by Harold Clark, Pp.37-38).

It was easy for him to reach this conclusion. For twenty-five years he carried on breeding experiments with moths and no matter how they varied, he still ended up with gypsy moths. His experiments finally indicated to him, variation is not evolution in the strict sense. Evolutionary biologists have recognized microevolution expresses the idea of change within the major groups and macroevolution expresses the idea of the origin of the higher categories, the family, order, class and phylum. This is why creationists, with evidence on their side, object to the word evolution in the "great" sense. They do not observe this phenomenon in the living world of animals or in the dead world of animals fossilized.

Hank Hanegraaff has given us the definitions of the differences between macro and micro:

"Macroevolution refers to large-scale changes – where one species transforms into another completely different species. For example, birds are said to have evolved from dinosaurs. This process would require the addition of new information to the genetic code.

"Microevolution refers to changes in the gene expressions of a given type of organism but does not produce a completely different species. For example, through selective breeding, dogs ranging from Great Danes to Chihuahuas have been produced from wolves. This process ... does not require new information because the changes are a function of the genetic makeup already present in the gene pool of the species."

The Farce of Evolution

P.172

What do pigeons and bulldogs have in common with other animals? Simply this, they have ability to change but that change is limited. Pigeons and bulldogs have variation in common with other existing groups of animals. Darwin selected animals with particular traits and allowed them to reproduce. He apparently *did not discover* what other breeders learned over the centuries – change produced in animals is limited and not open-ended. Although breeders can create rich diversity within "kinds" of animals, the "great changes" demanded by Darwinian evolution will never be observed. A pigeon will always remain a bird and a bulldog will always remain a dog. Darwin's theory of one type of organism being transformed into another type of organism will always remain a theory and never become one of nature's fixed laws.

CHAPTER TEN

I'LL BE A MONKEY'S UNCLE OR-
MY GREAT GRANDFATHER OF
A HALF MILLION GENERATIONS AGO-
WAS AN APE

In the *National Geographic,* November 2004, pages 7, 21, and 32 contain photographs of the orangutan and chimpanzee. David Quammen maintains that man has similar anatomy to apes and monkeys, allegedly proving that they share a common ancestor in their evolutionary history.

Comments:-According to many evolutionists, the chimpanzee is the closest kin to man. Therefore, the following question is to be considered:

DO EVOLUTIONISTS TEACH THAT MAN CAME FROM APES?

The *National Geographic,* over the course of the last ten years, has contended this sordid evolutionary doctrine to be true. Everything possible has been done to present evolutionary scholarship to the public in the form of graphic pictures by the best artists, colorful illustrations that are explicit and well-written articles by those authors who have been engrossed in their evolutionary themes. And what is the reason? Evolution is losing its hold on the public. More professionals are turning from belief in evolution to belief in creationism. Later, this fact will be confirmed when examples are presented.

There have been times when I asked people, why do you believe in evolution? Most people are not well enough informed to discuss their reasons from natural selection, macroevolution, slow progression of species within the fossil record, classification systems or cladistics, the "traditional" view of evolution, the "new synthesis" of evolution, ontogeny recapitulates phylogeny, Neo-Darwinism, morphology, convergent evolution, biogeography, population genetics, molecular biology, vestigial structures, or from punctuated equilibrium. Only one person in answering my question mentioned punctuated equilibrium as an explanation for her belief in evolution. But with further questioning, she hesitated and was reluctant to carry on the discussion [I do not intend to suggest that I am proficient in knowing all there is to know about the above reasons that evolutionists advance to back up their beliefs. I have studied evolutionism and creationism for a period that covers over fifty years but I am far from being a know-it-all] So, what proof of evolution do most people mention to support their viewpoint? When confronted with my question, almost every person answered in the following manner: "Well look at prehistory! The 'ape-men' prove evolution. Man came from the apes; that's proof enough. Do you ever look at the 'National Geographic' magazine? There is plenty of evidence in there; what else do you need?" Apparently, the *National Geographic* has served its purpose in attracting the public with its irreverent doctrine of man coming from the apes.

Notwithstanding, more and more people are becoming enlightened and drawing away from the evolutionary doctrine supporting the ape to man idea. Especially through books that are coming off the printing press from more than the publishing houses of so-called "right-wing extremists." For examples:

* Phillip E. Johnson is a graduate of Harvard and the University of Chicago. He has taught law for over twenty years at the University of California, Berkeley and courageously uses his legal knowledge in combating Darwinism. He wrote the books -

Darwin on Trial –evolution is based on faith, not fact.

Defeating Darwinism – by opening minds to the truth.

The Wedge of Truth – splitting the foundations of Naturalism.

Reason in the Balance - the case against naturalism in Science, Law, and Education.

* Jack Cuozzo is an orthodontist who spent 20 years of his life researching Neanderthal man, first hand in world museums. He wrote the book -

Buried Alive – This book presents the truth of how evolutionists "buried alive" the real history of the Neanderthal family.

Incidentally, Cuozzo, as a young man, was a product of two universities and had negative beliefs concerning the Bible and God as Creator. Later in life, after his study of the facts, he began to look away from false-science and the many contradictions in the doctrine of evolution. He claimed that it was the amazing grace of God that attended him in the writing of his book.

*Michael J. Behe is professor of chemistry at Lehigh University. He wrote the book -

Darwin's Black Box – This work is the biochemical challenge to evolution. Behe makes an overwhelming case against Darwinism on the biochemical level. He does not espouse creationism but he does point out that Darwin had no explanation for origins. He also demonstrates that life did not arise from non-living matter.

*Nicholas Comninellis, M.D. He wrote the book -

Creative Defense – This work presents a host of evidence against evolution. An incredible resource book wherein the author writes up documented evidence against the bad science of evolution. The author combats the inconsistencies of the evolutionary theory.

*Robert V. Gentry is a tenacious, courageous, scientist whose research produced remarkable documentation for creation. His research involved working with radioactive halos in ancient material and eventually produced the following technical book ...

Creation's Tiny Mystery – an account of a scientist's lonely struggle to prevent the scientific "establishment" from suppressing his gathered information.

Special Note: - The Foreword to this book was written by W. Scot Morrow, Ph.D., an associate professor of chemistry at Wofford College. Morrow is an agnostic evolutionist. He makes an appeal to fellow evolutionists and all scientists to read this book by a fundamentalist Christian and profit from its reading. Morrow states that "Robert V. Gentry is a scientist in the tradition of Galileo."

* James Perloff – is a free-lance writer based in Boston, Massachusetts. He was a contributing editor to the *New America* for several years; a former fanatical atheist and anti-creationist, understands the other side's point of view. He wrote the book -

Tornado in a Junkyard – a refutation of the relentless myth of Darwinism. It is a scientific case against Darwinism informally written for laymen. It is entertaining as well as educational. A warning label is affixed: THIS BOOK MAY CHANGE YOUR LIFE

And there you have it! - A short listing of books; mighty in their contents; written by impressionable intellects with God in their hearts. In spite of the rose-colored lenses worn by David Quammen, "polls show that less than 10 percent of the American public believes in the official scientific orthodoxy, which is that humans (and other living things) were created by a materialistic process in which God played no part."

Defeating Darwinism

Phillip E. Johnson

p.10

Before I move on, a word for *National Geographic*. My sister and brother-in-law have made certain that I received this magazine for the past 30 years. What other magazine can provide information of places and events from all over the globe? Even the articles on evolution keep me informed of recent discoveries and the thinking of its advocates. Time was, when evolutionists did not care to advertise that man came from apes or monkeys for fear of public opinion. But this was in spite of the fact: any laymen could have read Darwin's *The Origin of Species*. Did Darwin claim that we came from monkeys?

The following statement will speak for itself and the readers can decide:

"... Old World monkeys, all of which are characterized ... by the peculiar structure of their nostrils, and by having four premolars in each jaw ... Now man unquestionably belongs in his dentition, in the structure of his nostrils, and some other respects to the Catarhine or Old World division ... there can, consequently, hardly be a doubt that man is an off-shoot from the Old World Simian stem; and that under a genealogical point of view he must be classed with the Catarhine division."[Man, claims Darwin, is an off-shoot of the Old World Monkeys "however much the conclusion may revolt our pride."]

P.519

In the next statement, Darwin helps us swallow our pride:

"But we must not fall into the error or supposing that the early progenitors of the whole Simian stock, including man, was identical with, or even closely resembled, any existing ape or monkey."

P.520

I certainly feel better now. My pride was hurt in knowing that I came from Old World monkeys but in that I probably take after my ancestors (going back farther in time) who weren't as ugly as Old World or present day apes and monkeys, my pride is back on the mend.

But, Darwin's comments under the title – "On the Birthplace and Antiquity of Man"- once again "may revolt our pride." Darwin claims that:

"It is therefore probable that Africa was formerly inhabited by extinct apes closely allied to the gorilla and chimpanzee; and as these two species are now men's nearest allies, it is somewhat more probable that our early progenitors lived on the African Continent than they lived elsewhere."

P.520

Most evolutionists presently believe that man is closely related to the gorilla and chimpanzee. They believe that a HOMINOID ANCESTOR produced APE-LIKE CREATURES that SPLIT INTO SUBPOPULATIONS. The NEAREST KIN TO MAN is the CHIMPANZEE and GORILLA. The GIBBON and ORANGUTAN are distant relatives.

With the above arrangement, being typical of the anthropologists' belief in man's evolutionary relationship with the primates plus Darwin's previous statements, it is reasonable to conclude that evolutionists teach that man did come from apes. What would be the reason otherwise for evolutionists speaking of our ancestors as "ape-men"?

Further collaboration for this conclusion is confirmed by Cronkite's remarks:

" ... Walter Cronkite, in the television premiere of 'Ape Man: the Story of Human Evolution' declared that monkeys were his 'newfound cousins.' Cronkite went on to say: 'If you go back far enough, we and the chimps share a common ancestor. My father's father's father's father, going back maybe a half million generations – about five million years – was an ape.'"

The Farce of Evolution

Hank Hanegraaff

Pp. 56-57

Richard Leakey is a self-made Anthropologist. He is the son of the late Luis and Mary Leakey and has continued his parents' pioneering field work in East Africa. He has been called the "organizing genius of modern paleoanthropology."

 In his book entitled "Origins" he writes:

"There is an inescapable and persistent element of excitement in the search for the origins of humanity. It affects everyone, professionals and non-professionals alike, because there appears to be a universal curiosity about our past, about how a thinking, feeling cultural being emerged from a primitive ape-like stock. What evolutionary circumstances molded that ancient ape into a tall, upright, highly intelligent creature that, through technology and determination has come to dominate the world?"

P.7

The evolutionists Walter Cronkite and Richard Leakey have their opinions but I will still choose mine:

"We came into existence not by mathematical chance but rather by the charge of a Supreme Mathematician; Not by a hominoid ancestor but rather by the creative act of the Ancient One;

- Not by survival of the fittest but rather by the divine fiat of a Creator-Savior who made us fit to survive;

- Not by genetic mutations but rather by the machinations of the Mighty Progenitor;

- Not by orthogenesis but rather by the Originator of every genesis;

- Not as the offspring of gorillas but rather as the children of God;

- Not as relatives of monkeys but rather as sons and daughters of the Divine Maker;

- Finally, I came into existence not as the creation from the cladistic, evolutionary branch of animals but as God's unique crowning act of creation."

Richard Leakey has entitled his book *Origins* but there is no origin from the evolutionary standpoint for man's original ancestor. On all phylogenitic charts for the hominoids, there is a block labeled "hominoid ancestor" at the very top of the scheme for man's origins. Duane T. Gish sites the quote of Lowenstein and Zihlman in his *EVOLUTION: the Fossils STILL say NO!*:

"Imaginations run riot in conjuring up an image of our most ancient ancestor – the creature that gave rise to both apes and humans. This ancestor is not apparent in ape or human anatomy or in the fossil record, but is evident only in the unseen world of the genome within the cell."

P.212

The original ancestor of man and apes exists purely in the imagination of the mind of the evolutionist – it doesn't exist except through faith. Most phylogenitic charts printed in books on evolution are filled with broken lines. This means, nothing in the fossil record indicates relationships between one species and the next. The charts calls to mind Darwin's

Tree of Life - the branches are not affixed to the trunk because there is no trunk. More than this, there is no connection between twigs and branches. The charts, with their broken lines, indicate the exact same thing as does The Tree of Life...NO CONNECTING LINKS FOR COMMON ANCESTORS AND ESPECIALLY, MAN AND APES.

Paleoanthropologists use the term "Missing link" for every fossil-man or portions of fossil-men discovered. This is truly a subjective term. There are so many broken lines in their APE TO MAN charts; each new discovery has to fit in someplace. Where and how is up to the discoverer. The thinking is so subjective, each evolutionist has his own concept of the ape-man relationship and this, in turn, calls for a new chart. Anyone with the slightest degree of thinking power can easily discern that these imaginatory charts have little or nothing to do with science.

Although I feel that men and women like the Leakeys, have wasted their lives in investigating the illusory tracking of man's ancestry; in drawing up chimerical charts; in conceptualizing that each discovery is a missing link; in fabricating whole lines in place of broken lines; in turning to a fictitious brand of science; nevertheless, I have compassion in my heart towards them. In my imagination, I see such men and women roaming the hills and deserts of far away places; suffering the discomforts of extreme heat and cold; laboring under the most adverse conditions; avoiding stinging scorpions, poisonous snakes, and biting insects; giving up the luxury of a cold refrigerator, a hot shower, and a warm bed. Field work is not for the faint hearted or for that individual unaccustomed to arduous work. But I see those same men and women driven by an inward force which they themselves do not understand; inspired by the sunrise of each new day and enraptured by the sunset of each fading day; delighting to the invitation of the natural world; responding to the thrill of each fossil unearthed.

I often ask God to seek out such fossil hunters and that they (in their nightly dreams) would hear Him softly whisper to His Son, "Let us make man in our image, in our likeness" and as they daily dig for fossils, they envision within their mind the words of David etched in the sand, "For you created my inmost being; ... I praise you because I am fearfully and wonderfully made ..." (Genesis 1:26; Psalm 139:14).

Darwin wrote:

"... Two great stems, the New World and Old World monkeys; and from the latter, at a remote period, Man, the wonder and glory of the Universe, proceeded ... ... Unless we willfully close ours eyes, we may, with our present knowledge, approximately recognize our parentage; nor need we feel ashamed of it…The most humble organism is something much higher than the inorganic dust under our feet; and no one with an unbiased mind can study any living creature, however humble, without being struck with enthusiasm at its marvelous structure and properties."

The Origin of Species

P.528

I can picture Darwin's readers, with tears streaming down their cheeks, as they respond verbally to this emotional and gushing Darwinian plea: "Yes, it's true, all of it" – they say – "I am not ashamed of this poor, little monkey … this humble organism that brought the glorified me into existence. Yes, it's true (between sniffs), the monkey is my father."

I agree with Darwin, but only on his last sentence in the above declaration. I am enthused or impressed with the marvelous structure and properties of a humble, living creature … but only in the context of its coming forth from the hand of the Creator. I stand in awe, in contemplating the miracle of every humble beast. However, I marvel more at man; man created in the image of God with complicated structures; man who is distinct from all other creatures; man who is not the son of "Cheetah" but rather the son of God; man who is not the brother of the monkey but the brother of Christ. I marvel at man who was given the power of speech and is able to say, "I will praise thee; for I am fearfully and wonderfully made: marvelous are thy works; and that my soul knoweth right well."

Ps. 139:14

I also marvel even more at evolutionists who admire the marvelous structures and properties in monkeys yet deny the Creator of monkeys and who would rather attribute their sonship to that of an ape rather than to the person of God.

CHAPTER ELEVEN

THE ORANGUTAN'S VISIT TO DODGER STADIUM

In the *National Geographic*, November 2004, page 21, David Quammen attempts to make a case for evolution through morphology:

SIMILARITIES OF ANATOMY IMPLY COMMON ORIGINS.THE ORANGUTAN HAS LONG ARMS, BUT ITS PAIRED FOREARM BONES RESEMBLE THE RADIUS AND ULNA IN A HUMAN. THE ORANGUTAN HAND IS SO SIMILAR TO OURS THAT IT MIGHT FIT IN A FIRST BASEMAN'S MITT.

Comments: - "The orangutan hand is so similar to ours that it might fit in a first baseman's mitt." These remarks by David Quammen are meant to be a witticism but they fall short of producing a laugh from those of us who know that, once again, Mr. Quammen is attempting to hoodwink the public with one of the four main proofs (?) of evolution – MORPHOLOGY.

Morphology is that science which teaches comparative anatomy – the analyzation of the comparable structures in various animals and plants. This is a good thing as far as it goes. When it comes to animals, scientists have a fairly good idea how to construct scattered bones from the past by comparing them with articulated skeletons of the present. The paleontologist can figure out muscle attachments and the general body shape of extinct animals. Comparative anatomy can also facilitate the biological study of fossils through their classification.

But when comparative anatomy is used as a method to teach the history of animals – such as evolutionary relationships and lines of descent – it falls short and removes itself from objective science to subjective philosophy. Deceptive science leads men to think that because certain biological forms have similar structures, they must have a common evolutionary ancestor. How can the evolutionist use common patterns of organization as a proof positive of evolution? The creationist has just as good, if not better, an argument that anatomical patterns and designs have their common origin in a Creator who fashioned anatomy in the first place.

Let us return to Quammen's remark about the hand of the orangutan being similar to ours and might fit into a first baseman's mitt. Let us carry this witticism through to its logical conclusion. Consider taking Mr. Orangutan down to Dodger Stadium to have him try out for the Dodger's first baseman position. Because he is quite different from other ball players on the field, we might imagine our conversation with the team manager of the Dodgers to be something like this:

US — "Mr. Orangutan is here to try out for your team as first baseman."

Team Manager — "We can't use him."

US —"Why not?"

T.M. —"All my players are Homo-sapiens ... 'wise men.' Your orangutan is literally a 'man of the forest.'"

US — "Our player is kind of like yours. His hand will fit into a first baseman's mitt."

T.M. — "What does that prove? Can he hit and field?"

US — "It proves that he is a little like your players. He has five-digital hands to prove it."

T.M. — "He is too difficult to outfit."

US — "What does that mean?"

T.M. —"Well, for example, consider his head. Where would we find a cap to fit him? I understand that he has a cranial capacity of 411 cubic centimeters. Most of our guys reach 1350. Consider next, a base ball jersey. All our players have shorter arms than legs. Your player's arms are too long. Why they even touch the ground. I saw him walking on them. Isn't that why they call him a Quadrumana? – Because he can walk on all fours. All our players are built for walking or running. And that brings me to another item... How can your player field ground balls and run to first base when the time comes?"

US — "You don't have to make fun of him."

T.M. —"I'm not making fun of him; just pointing out some main differences. Let's talk about outfitting him with baseball shoes. All our players have their feet constructed for running. Look at Mr. Orangutan's feet! He doesn't have a well-formed arch and his great toe isn't in line with the other toes. How can he run for a fly ball on a pop-up? And how can he reach up for the coming ball on an over-throw to first base? With that curvature in his spine and his head not perfectly balanced on his spinal column, he will make too many errors."

US   – "You are going too far! At least his body weight is a nice lean 121 pounds and he can really travel the bases."

T.M.   – "Who are you kidding? I've seen him walk and run on his way into the stadium. He raises his arms up, runs about two feet, and down again to all fours. And with that flimsy weight, he should be able to drive the ball far into the seats in center field – I think not."

US   – "This time, you have really gone too far!"

T.M.   – "I'm just beginning to warm up. What is more, I don't think he is intelligent enough to play the game."

US   – "Don't you know that the total volume of the brain compared to the body is not an absolute criterion of intelligence?"

T.M.   – "But between different species it is a strong indicator."

US   – "You have an answer for everything."

T.M.   – "That's why I am the team manager."

US   – "We think Mr. Orangutan will make your guys laugh when they are off the field and socializing."

T.M.   – "I don't know. For a ball team, my guys are very aesthetic."

US   – "What do you mean?"

T.M.   –"I don't think Mr. Orangutan would be affable for my ball team because he has no ability to grasp abstract ideas like truth and beauty. Does he like good music? ; show an interest in poetry; read the classics; appreciate the work of great artists; and does he ever go to church and demonstrate an interest in spiritual matters?

US   – "He never let us know by his 'signs'."

T.M.   – "I'll tell you what! Why don't you come back another day? Oh, and bring your chimp friend with you! I understand that he is smarter than Mr. Orangutan, his distant cousin."

For every comparison that the evolutionist makes between apes and man or between monkeys and man, the creationist will point to more differences. There are at least 50 major differences between man and the anthropoids. While it is true that similarities exist, the differences need to be taken into account. An old statement (but still true) from the pen of an evolutionary zoologist, Austin H. Clark:

"Man is not an ape, and in spite of the similarity between them there is not the slightest evidence that man is descended from an ape."*(The New Evolution: Zoogenesis)*

One similarity that must at this point be discussed and not passed over is the similarity of human and chimp DNA. This is the traditional argument presented by evolutionists. The DNA serves as evidence to prove that there is an evolutionary relationship between humans and primates.

But first, I have some commentary about Linnaeus (1707-1778). Linnaeus is my favorite creationist of scientific history although, according to his autobiographical documents, there is some doubt about his humility. One might say Linnaeus was speaking not from the standpoint of his ego but from the perspective of heart-felt conviction. He wrote, writing in the third person, "God has permitted him to see more of His created work than any mortal before him." Of course Linnaeus was referring to his tremendous work on the classification of plants and animals. Some commentators have recast Linnaeus as being close to evolutionists but that comes from a misunderstanding of his writings.

The Swedes also liked Linnaeus – their home prodigy – by placing him on two of their bank notes. The evolutionists like Linnaeus because they could not conceive of Darwin without Linnaeus; of evolution without previous knowledge of the method and practice of taxonomic arrangement; the creationists like him because he remained a creationist all his life and believed that God brought life into existence. However, an examination of lists of species in *Systema Naturae* would differ with many creationists because of the way Linnaeus divided up many of the Genesis kind into "species." For example, he believed that horses, asses, and zebras were different Genesis kinds. Many special creationists would differ, believing them to be of the same "kind" of animal but demonstrating varieties. Linnaeus did this same unexplainable thing in the naming of plants. For example, he gave separate species names to spring wheat and winter wheat which creationists would recognize as one kind or species. Linnaeus wasn't perfect in his classification system and neither is any other taxonomist since his time. Linnaeus did not have advanced information and neither does any other scientist.

There are reasons for mentioning Linnaeus. On the one hand, Darwin did not invent the classification system; Linnaeus was the father of modern taxonomy. On the other hand, our modern taxonomic system is a suitable way of indicating the similar or dissimilar morphology of plants and animals. Such charts of classification were not meant to provide scientists with a picture of common ancestral or blood relationships. Linnaeus was a creationist and he never intended for his charts to have this meaning.

Such visual aids were taken over by the evolutionists who attributed to animal classifications the extended definition of blood relationships. Linnaeus included humans with the apes in his classification arrangement and scientists, before the time of Darwin, were very much aware that humans and anthropoid apes are physically alike. But Darwin, along with other evolutionists, superimposed their philosophy over the empirical observations of Linnaeus. But this old philosophy of blood relationship between animals lacked scientific proof.

However, neo-Darwinists have made the bold contention that these blood relationships are now evidenced through the study of genetics. Do genetics fill in the gaps of the classification systems? Classification is merely a guess concerning species. The Linnaeus guess at relationships of animals and plants was often a poor one at best. The evolutionist's guess at species, from a "scientific" standpoint, was even a poorer guess. Classification systems only indicate the similarities and dissimilarities of animals. However, the evolutionist supposes that genetics confirms his evolutionary position of blood relationships for the various organisms. For example, since the human and chimp DNA are similar, it is reasoned that people are nothing more or nothing less than just an evolved ape. But is this evolutionary teaching valid? Will it stand up in the Empirical Court House of True Science?

Let us begin our short but significant investigation. At first, evolutionists put a lot of stock in morphology or comparative anatomy. How does the animal look in its outward features? This was an important question for the reason that similar forms and shapes were supposed to show sequential relationships. One class of animals was imagined to be linked up with another class by a linear series of transitional forms. The evolutionist believed that all present forms of life evolved from simple forms. But this concept was highly speculative. Many evolutionists were aware of the fact that the fossil record (past or present) bore no real evidence that there ever was a continuous series of forms from the most simple to the most complex. The scientists worked from the standpoint of logic but one biological fact – the actual discontinuity of animals that exist in nature – is enough to cause logic to succumb to the truth of reality

The evolutionists turned to another branch of study called comparative biology. In the Classification Systems initiated by Linnaeus, evolutionists could now assess the differences between species by comparing their structure at a *gross morphological level*. That is, not only comparing the forms and outward features of animals but their total life make-up – their biology. Comparative biology was how the animal functions. How and what does he eat, drink, and breathe? Michael Denton claims that "comparative biology was no more or less than comparative anatomy." Comparative biology still did not furnish evidence that classes of animals were linked by transitional forms or that they had a common ancestor. There was no chain of organisms which could be traced back to a common blood ancestor. The creationist could sit down with the evolutionist in front of a museum display case and reason: "I don't believe that these animal forms evolved from a common ancestor. I believe their design and symmetry demonstrate that they come forth from the hand of an Intelligent Creator who designed and spoke them into existence. I believe God made plants and animals with the same key molecules. Therefore, I would expect man and other organisms to drink the same water, eat similar foods and take the same air into their lungs." On the other hand, the evolutionist argues from the above mentioned point of view. The conclusions are in juxtaposition but both make sense. Both opinions are logical but the evidence is subjective.

Comparative biology can be easily linked with the biochemical similarities in animals because what we eat, drink, or the way we breathe depends upon our metabolism. That is why the creationist in the above match-up, reasoned from the standpoint of biochemistry that because man was made up of certain chemical constituents similar to the animals, man and animals would be expected to respond the same way to their environment. But the evolutionist wants to take this concept further. He wants society to believe that the macromolecules that exist in man can be traced back to remote ancestors of the past and that these macromolecules are the results of evolutionary change by mutations that have been accumulated throughout the scope of time. He asserts that these macromolecules appear in similar life forms to show that they have evolved. He makes the bold assumption that biochemistry determines the patterns of biological relationships – that the various sequences of proteins should be arranged into an evolutionary series.

The evolutionist believes that where *comparative anatomy and comparative biology* have failed because of the many gaps in the fossil record, *comparative biochemistry* will provide the confirmation of evolutionary sequence of organisms and provide the connecting links that have eluded paleontologists for one hundred and forty years since the time of Darwin. The evolutionist has also made the notorious charge - stemming from his overconfidence – that because humans and chimps have similar DNA, this must be indicative of evolutionary relationships. Is the blood relationship between the human and the chimp based on actual fact?

Before we answer this question, a few facts about biochemistry are in order. Biochemistry is one of the most difficult subjects on earth – next to Astrophysics. From the writings of these men (evolutionists as well as creationists) one can extrapolate the numerous inconsistencies of the molecular evolutionary perspective. Evolutionists have practically given up on the fossil record and turn to molecular biology for their answers. The claim is made that evolutionists no longer have need of turning to the fossil record to establish the interconnecting link of species. *The superfluous belief is that molecular biology will establish the essential links.*

In response to this assertion and to point out the first of many problems for the evolutionist as he turns to the field of molecular biology, I go to the words of Michael Denton (a non-creationist):

"However, as more protein sequences began to accumulate during the 1960s, it became increasingly apparent that the molecules were not going to provide any evidence of sequential arrangements in nature, but were going to reaffirm the traditional view that the system of nature conforms fundamentally to a highly ordered hierarchic scheme from which all direct evidence for evolution is emphatically absent. Moreover, the divisions turned out to be more mathematically perfect than even most die-hard topologist would have predicted."

Evolution: A Theory in Crisis

Pp.277-278

This observation by Denton who is a molecular biologist and is best known for his research, is contradictory to the previous assertion that this new "scientific" approach will make up for the lack of transitional species. This bad news, no doubt, disappointed proponents of the Evolutionary Doctrine.

A number of problems equated with molecular studies are now listed for your consideration:

• There are no fossil molecules. Molecular biology only deals with living organisms and their molecules. Past life can't be observed because there are no fossil molecules. How can you analyze that which has no existence? Evolutionists are worse off than they were with the fossil record.

• The "Molecular Clock" has too many assumptions: One "sets" the clock by making the assumption the times when certain organisms first evolved, or when two organisms diverged. First of all, one has to assume evolution to be a fact; secondly, the time element of divergence also must be calibrated *on an assumption* for there is no other way. The molecular clock still relies on a fossil record without molecules. The only molecules that exist in the fossil record stem from the imagination of evolutionists. This is not true empirical science but rather "science fiction."

• When the Cytochrome Cs … … bacteria were compared, the data was highly contradictory to evolutionary predictions. The data recorded into a Table of so-called facts, had no predictive value. The reason that it has no predictive value is explained by Philip E. Johnson in the finale of this list.

• Molecular data often indicates the opposite of predicted evolutionary charts … Example: Data shows that amphibians are much more distant relatives to mammals and birds than are the reptiles.

• When molecular data of different proteins are compared, it is not possible to erect a single phylogenitic tree. Many trees emerge and these groupings contradict the supposed evolutionary relationships which are supposed to be progressive.

• Evolutionists, Christian Schwabe, Gregory Warr, and non-creationist, Michael Denton, have all attacked the idea that amino acid sequence data can be used to construct evolutionary phylogenitic trees.

• The only data for molecular differences is found at the tips of the branches of the so-called evolutionary tree. All the other information is filled in with speculation. Again, the *public is hoodwinked* into not recognizing that the numbers assigned to ancestral proteins are not numbers derived at by actual investigation of molecules but based on the assumption monophyletic evolution has occurred. As Gish writes: "The proof of evolution thus is merely the assumption of evolution." … … … Enough problems have been cited to show that "molecular evidence" may be more of a problem for evolutionists than a help. (The above seven problems have been extrapolated from Duane T. Gish's book - *Creation Scientists Answer Their Critics* and Pp. 275-294)

I have one more problem for scientists who study molecules with an evolutionary bias. This problem is taken from the book, *Darwin On Trial* by Phillip E. Johnson and page 94: Cytochrome C comparisons place every plant and animal species in approximately the same molecular distance from any bacterial species. The problem is obvious. Where are the so-called intermediate organisms? If molecules evolved gradually then there must be intermediates to bridge the gap. The missing gaps are not only conspicuous through molecular studies but the same studies reveal that there "are a greater number of fundamental divisions in the living world that had previously been recognized."

Regardless, most evolutionists claim that man's nearest kin among the apes is the chimpanzee. Of course there are many statements such as the following that expresses what other evolutionists think about ape and human relationships. David Pilbeam claims:

"There is no clear-cut and inexorable pathway from ape to human being."

Human Nature (June 1978)

P.44

"I know that, at least in paleoanthropology, data are still so sparse that theory heavily influences interpretations. Theories have, in the past, clearly reflected our current ideologies of the actual data"

P.45

We should understand that these statements were made almost two decades after biochemistry became a popular way of comparing organisms. Let us check the "actual data" to see if the chimp is a close relative of man. When Pilbeam claims that there is no clear-cut pathway from ape to human, the pathway *must include molecular studies* because molecular studies preceded the Pilbeam claim by 20 years. The inference being, of course, that molecular studies *have not cleared the pathway* as predicted by evolutionary scientists.

In spite of this, chimps and human beings are affirmed by evolutionists to have similar DNA ranging from figures of 97 percent up to 99 percent. The public, unfortunately, believes this information is hot news just off the press. Publications keep on repeating this 25 year old argument and the public persists in latching onto it because it neglects to investigate the significance of such a matter on its own. When two organisms are alike in form and shape (comparative anatomy) and are adapted biologically (comparative biology) to similar environments in the food they eat and the air they breathe in, creationists would expect chimps and humans to have similar DNA. The creationist sees these similarities as not being problematic because he understands their significance.

It is a fact that evolutionists have over-emphasized the similarities without giving proper emphasis to the dissimilarities. Dr. Allen is a former senior lecturer in genetics in South Africa. He holds a Ph.D. in genetics from the University of Edinburgh, Scotland. Let us observe what he claims is the significance of molecule studies between a chimp and human being. Dr. Allen cites Cytochrome-c, a protein and gene product which seems to occur in practically all living organisms. Its role is to function as a key enzyme in oxidation reactions. Molecular biologists have claimed that the similarity of Cytochrome-c taken from humans and chimpanzees leads to the conclusion that these two organisms are closely related.

But Dr. Allen writes:

"Apart from the single gene controlling the constitution of Cytochrome-c, humans and chimpanzees differ in many thousands of other genes. As a conservative estimate, let us say 5,000. What the theory of evolution is saying is that while humans and chimpanzees have evolved independently from a common ancestor so as to now differ in these 5,000 genes, there has been no change in the 93 amino-acids specified by the Cytochrome-c gene, and this in spite of there being no functional constraint on change in any of the latter. I find this to be an unacceptable claim."

(The James S. Allan quote is taken from page 130 in the Book *In Six Days* edited by John F. Ashton.)

The Answers Book claims that because chimps and humans have morphological similarities "We would expect similarities in their DNA. Of all the animals, chimps are most like humans, *so we would expect that their DNA would* be *most like human DNA, but not totally like human DNA."*

Page 115 [Emphasis mine]

Also, a page later, 116, "The human DNA has over 3 billion nucleotides. *Neither the human nor the chimp DNA has been anywhere near fully sequenced to allow a proper comparison."* [Emphases mine]

This fact, within itself, should minimize the similar DNA dispute.

The percentages of similarity between chimps and humans do not mean they "evolved" from a common ancestor. Even the low percentages of the base pairs of DNA left over (the differences) from the high percentages of comparisons between the two organisms, would be too much for mutations or random changes to cross. Evolutionists claim that the passing of millions of years are necessary to have humans (from the assumed ancestor that they have in common with the chimpanzee) to be molded into "modern man" by the process of mutation, genetic drift, and natural selection.

Dr. Allen writes in *Six Days* and P. 131:

"When I consider mutation rates, the 'cost' of the substitution of each new mutant gene in a population in terms of the number of 'genetic deaths,' the assumed number of mutant gene differences between evolutionary stages, and the population size necessary to accommodate such a large number of successive mutations, I find that there is a remarkable lack of evidence for the 'evolution of man.'"

He also considers the implications that 150,000,000,000 forerunners of "modern man" needed to substitute one gene for another by natural selection is far too many "genetic deaths" to have such a shortage of evidence in the form of fossils, tools, or whatever for the existence of cave-dwelling hunters and so-called ape-like pre-humans.

This section has stressed the scientific perils that investigators may find themselves falling into when they persist in making comparisons between humans and apes. The weaknesses have been pointed out in the morphological or anatomical comparisons, the biological comparisons, and the biochemical comparisons.

Solly Zuckerman in *Beyond the Ivory Tower*, Pp.19 and 64, makes these distressing observations about the human reasoning powers of scientists:

"... the interpretation of man's fossil history, where to the faithful anything is possible – and where the ardent believer is sometimes able to believe several contradictory things at the same time.

".....no scientist could logically dispute the proposition that man ... evolved from some ape-like creature ... without leaving any fossil traces of the steps of the transformation."

No scientist *could* but a lot of scientists *do* accept the theory that men evolved from apes *in spite of the contrary evidence derived from all areas of investigation.*

Michael Denton, in ending his marvelous chapter on the biochemical comparisons of various organisms, complements Zuckerman's observation with his own revealing statement on the sad state of men's thinking when it comes to the subject of evolution:

"Despite the fact that no convincing explanation of how random evolutionary processes could have resulted in such an ordered pattern of diversity, the idea of uniform rates of evolution is presented in the literature as if it were an empirical discovery. *The hold of the evolutionary paradigm is so powerful that an idea which is more like a principle of medieval astrology than a serious twentieth-century scientific theory has become a reality of evolutionary biologists.*"

Evolution: A Theory in Crisis, P.306 [Emphasis mine]

Many individuals confidently thinking they are abreast with the times are duped into believing that they have been introduced to the latest scientific news release. They are unknowing that such news probably was derived from an old newspaper kept in the files and already discolored with age. The "news" was erroneous *then*; the "news" is erroneous *now* – MAN HAS NO BLOOD RELATIONSHIP TO THE CHIMPANZEE OR ANY OTHER ANIMAL.

CHAPTER TWELVE

RATITES, RATITES, EVERYWHERE!

In the *National Geographic*, November 2004, pages 22-23, David Quammen asks the following question:

IN THE WILDS OF ARGENTINA DARWIN SAW TWO SPECIES OF LARGE FLIGHTLESS BIRDS; ONE OF THEM CALLED DARWIN'S RHEA IN HIS HONOR. WHY DID SOUTH AMERICA HARBOR THESE SIMILAR FORMS, RATHER THAN OSTRICHES, AS IN AFRICA, OR MAYBE MOAS, AS IN NEW ZEALAND?

Comments: - *Like Darwin*, creationists believe that animals were spread out over the globe from common centers; crossed ancient bridges; were carried over the seas and oceans on natural rafts; that were slow in mobility, traveled great distances with each successive generation; supplied the earth with abundant varieties,etc. The scientific name for animal distribution is called Biogeography.

Unlike Darwin who saw varieties of rheas and assumed their convergent existence had evolved from a common ancestor that was, in fact, a different animal on the phylogenitic tree, creationists believe that each basic type of animal retains its identity. Darwin not only had an obsession with progression but believed in variation without limitation. The creationists believe in variation of rheas but do not believe in unlimited variation that would change a rhea into some other form of life.

FACTS ABOUT SOME RATITES – FLIGHTLESS BIRDS (From *Webster's Family Encyclopedia* cited under "Sources"):

Extinct Moas are from New Zealand; had strong legs and were fast but not fast enough to outrun the Polynesian settlers who hunted them for food. In the words of Stephen Gould, "Who could resist a 500-pound chicken?"

The Kiwis are also from New Zealand; 10-16 inches long; having large claws; National Emblem of New Zealand; beautiful cloaks were once worn by chiefs made by Maori artisans from kiwi feathers. The nocturnal kiwis escaped the butchering of their larger moa relatives; no fossils except in the Pleistocene yet offering no light on origins.

The Emu is from Australia; Order: Dromaiidae; 60 inches long; 100 lbs; naked blue spot on each side of neck; can run 30 mph; no fossil record.

Rhea – Darwin saw this flightless bird of similar forms in South America - the wilds of Argentina; Order: Rheiformes; 47 inches long; (3) toes: no fossil record.

Ostriches are from Africa; Order: Struthioniformes; (2) toes; the only ratite to have a fossil record but the most this fossil record can tell paleontologists, is the uncertain connection that ostriches have with birds (Gish). The fossil record tells them nothing about the ancestry or origins of these *SPECIALIZED FLIGHTLESS BIRDS*. There is *NO LIGHT ON RATITES*. David Quammen should pay close attention to the reality of the ratite fossil record. These birds were highly specialized and it should be easy to find transitional forms leading up to them. That is, if evolution is true. However, there are no fossils to make the case for evolution.

Quammen mentions similar forms of rheas in the Argentina wilds of South America. On the one hand, Darwin came to believe these similarities were the result of evolution. On the other hand, the creationist reasonably assumes that rhea variation is the result of speciation. In the case of Argentina, a special kind of animal split from its parent population (ISOLATION) and eventually created diversity through drawing from the original gene pool (SPECIATION). We should note that diversity constitutes small changes within an animal kind resulting from a recombination of existing genes – not some major change that would produce a new creature altogether. Variation is not unlimited; speciation is not an indication of macroevolution.

Finally the question constructed by Quammen, at the beginning of this chapter, is almost as vague as the fossil record of the ratites. It is just another ploy to hoodwink the public into believing the evolutionary-convergence-spin of similar forms rather than similarities resulting from their adaptations to similar ecological conditions. For Darwin, rhea with its similar forms came to mean a present variation which, supposedly, was a strong indicator of past evolution from a common ancestor. Quammen is eager to point this out to the public. I am just as pleased to point out to my readers, the rheas left behind NO FOSSIL RECORD to attest to the alleged common ancestor. Once again, Quammen's evidence is based on presumption - not a good guarantee that evolution has taken place among the ratites.

CHAPTER THIRTEEN

THE ANTELOPE WHO WANTED TO
BECOME A WHALE –
A "WHALE" OF A LIE

In the *National Geographic*, November 2004, pages 24-25, David Quammen makes this hopeful but fantastic observation concerning the whale like creature, Dorudon:

AT A DIG IN EGYPT A TEAM OF PALEONTOLOGISTS, AMONG THEM THE UNIVERSITY OF MICHIGAN'S PHILIP GINGERICH, FOUND THE NEARLY COMPLETE SKELETON OF A WHALE LIKE CREATURE NOW CALLED DORUDON ... ... BUT ITS VERY EXISTENCE TESTIFIES TO THE WHALE'S DESCENT FROM A FOUR – LEGGED ANCESTOR.

Comments: - David Quammen met with the paleontologist, Philip D. Gingerich. How much paleontology did Quammen learn from the great master of ancient cetaceans? Whatever he learned about whales was overshadowed by Gingerich's "intellectual passion and solid expertise with one other trait that's valuable in a scientist: a willingness to admit when he's wrong."

According to Quammen's account, Gingerich was incorrect about a certain whale identification that Gingerich made as a result of his field trips to Egypt and Pakistan. Here is the report (in a nutshell) that Gingerich gave to Quammen. The reader can study the report (for more details) on page 31 of the *National Geographic*, November 2004. An attempt will now be made to give a synopsis of this outlandish and farfetched account of whale evolution:

In the late 1970s, Gingerich discovered Pakcetus in Pakistan. A former student of Gingerich, Hans Thewissen discovered Ambulocetus natans (walking-and-swimming whale). Gingerich and his team dug up several more whales, including, Rodhocetus balochistanensis. After these first discoveries, Gingerich formulated his conclusion of the finds and Quammen reports:"Gingerich told me, he leaned toward believing that whales had descended from a group of carnivorous Eocene mammals known as mesonychids." Gish in his *EVOLUTION: The Fossils STILL say NO!* , writes: "The mesonychids were wolf-like, hoofed carnivores that as far as anyone knows, never went near water except to drink." P.200

What is the most amazing part of this account, Gingerich not only believed in his fantastic and farcical conclusion but most paleontologists, by the end of the 1990s, agreed with him. We might add, as Quammen dutifully and eagerly scratched Gingerich's information into his notebook he, as well, believed this paleotological nonsense. There is a great deal of information that had been left out of Quammen's report. For example, the truth about Gingerich's first "whale" digs in Pakistan. In 1983, newspaper headlines announced the discovery of a so-called primitive whale, Pakicetus inaclus, as having a land-mammal ancestor, the carnivore, Mesonyx. Duane Gish questions this whale (?) discovery. He gives seven crucial reasons why in his *EVOLUTION; the fossils STILL say NO!* P.201. [This is a shortening of his account]

1.   The fossil material was found in a formation that was land deposited.

2.   The fossil remains associated with Pakicetus are dominated by land animals.

3.   The evidence of nonmammalian remains such as snails, fishes, turtles, and crocodiles   indicate an environment not expected for a whale or whale-like creatures.

4.   The auditory mechanism of Pakicetus was that of a land mammal, rather than that of a whale.

5.   There is no evidence to confirm that Pakicetus could hear, as a whale hears, directionally under water.

6.   There is no evidence of vascularization of the middle ear, as in a whale, to maintain pressure during diving.

7.   The teeth resemble those of the mesonychids, which possibly fed on carrion, mollusks, or tough vegetable matter.

Gish says, "On the basis of this evidence, it seems most likely that Pakicetus was nothing more than a land mammal with no relationship to marine mammals."

But Gingerich, the master paleontologist, who is an expert on ancient cetaceans, tells us otherwise. When Gingerich works at a dig and finds fossil remains in a terrestrial deposit; in association with mostly land animals; in an environment not expected for a whale; in a state of having its auditory mechanism be that of a land mammal; as having no evidence of a whales ability to hear directionally under water and to maintain pressure during diving; last of all to have wolf-like teeth; then Gingerich must assume that the fossil remains represent a marine mammal which is family-connected to a land mammal. So much for Gingerich's "solid expertise" that Quammen assures us to be one of Gingerich's valuable traits of character.

Quammen should have taken the time to read Gish's extraordinary book, EVOLUTION: the fossils STILL say NO! And especially the section on "Marine Mammals." Gish refutes this whale (?) finding of Gingerich and his team and the so-called connections to the meat-eating mesonychids. Stay tuned "for the rest of the story!"

Quammen mentions the second trait of Gingerich – "A willingness to admit he's wrong." According to Quammen, this trait makes a valuable scientist. After Gingerich's fossil discoveries, molecular biologists suggested that the whales had descended from artiodactyls such as antelopes and hippos – not from meat-eating mesonychids. In the year 2000, Gingerich chose a new field site in Pakistan. In discovering a pulley-shaped anklebone "he had a moment of humbling recognition." Quammen writes, "Suddenly he realized how closely whales are related to antelopes." Yes, the molecular biologists were right, reasoned Gingerich – the whale had an antelope as its ancestor.

I scratched in my notes on pages 30-31 of the *National Geographic*, telling of Quammen's visit, these words ... ... "At first, the paleontologist Gingerich wanted us to believe that a wolf-like animal was drawn to the sea and later became a whale. Later, he was big enough to suddenly admit he was wrong; the antelope was drawn to the sea and Presto! ... The antelope is now a whale – a transitional form??? Anybody who can call an antelope a whale, believes in a *whale of a lie* for transitional forms in the fossil record."

[Readers think of how difficult it is to believe in the transition from an antelope to a whale, even if given considerable amounts of time! Gish describes the transition from a cow to a whale but evolutionists have cited so many different land animals turning into whales, it is hard for the creationist to keep up. However, think of a female antelope in place of a cow and Gish's words still apply]

Gish gives the following description:

"Let us stop to consider what this creature would look like halfway through this evolution. Her tail is only partway flukes......... and still has to spend much of her time on land, and it is not much help yet for swimming. Her front legs are surely a great embarrassment. They are getting shorter and shorter. Her front legs are surely a mess, too. They are halfway between ordinary feet, and legs, and flippers. This poor thing is at a tremendous disadvantage, while on land. She can't walk, or if she can, it would be terribly awkward. When she is in the water, she looks awfully silly trying to swim, with those part front feet – part flippers, and with a part tail – part flukes. Where does she have her babies, in the water or on the land? If she has them in the water, how does she keep them from drowning, and how does she nurse them? The whole idea that some hairy, four-legged mammal ventured into the water and gradually changed into a whale, during millions of years, is absurd. A whale is marvelously designed for life in the water."

The Amazing Story of Creation

Pp. 46-47

We must ask ourselves, how could any scheme in nature drive an antelope from her natural habitat into the water, a completely unnatural habitat for a four-legged mammal. We must perceive the inane stupidity associated with this completely illogical piece of so-called science on the origins of the whale. James Perloff, in his book *Tornado in a Junkyard*, written against the relentless myth of Darwinism, quotes from the British author Douglass Dewar, a fellow of the Zoological Society:-

"Both whales and sea cows swim by the up and down movement of the great flattened tail. Such movement is impossible in a land animal that has a pelvis, but a well-developed pelvis is essential to every land animal which uses its hind legs for walking ... I have repeatedly asked evolutionists to describe or draw the skeleton of a creature of which the pelvis and hind legs are anatomically midway between the state that prevails in whales and sea cows on the one hand, and a land quadruped on the other. No one has accepted the challenge, and of course a fossil of such a creature has not been found, and never will be."

P.15

In the comments on Darwin's Tree of Life, our thoughts were directed to Quammen's statement, "No one needs to, and no one should accept evolution merely as a matter of faith." Are we supposed to believe that evolutionists don't exercise even the smallest degree of faith when it concerns the transition from a four-legged mammal into a marine mammal, such as a whale, with its marvelous equipment for diving to great depths and being able to "see" with a sonar or echolocation system?

The eloquence of Duane Gish speaks for the faith of evolutionists which most of them deny having:

"It is clear that the evidence weighs heavily on the creationist side concerning the origin of marine mammals. It requires an enormous faith in miracles, where materialist philosophy actually forbids them, to believe that some hairy, four-legged mammal crawled into the water and gradually, over eons of time, gave rise to whales, dolphins, sea cows, seals, sea lions, walruses, and other marine mammals via thousands and thousands of random genetic errors. This blind hit and miss method supposedly generated the many specialized complex organs and structures without which these whales could not function, complex structures which in incipient stages would be totally useless and actually detrimental. Evolution theory is an incredible faith."

EVOLUTION the fossils STILL say NO!

Pp.207-208

Back to page 25 in the *National Geographic* and the comment of David Quammen in writing on the Dorudon whale, "But its very existence testifies to the whale's descent from a four-legged ancestor." Over this page I made my comments on homologous structures and genetic data. In this case, the homologous structures would be the five-fingered feet of the whale compared to the five digits in the hands and feet of most land mammals:

"Living specimens that have homologous structures do not testify of descent through genetic data; in fact genetic data contradicts descent. With no possibility of acquiring genetic information from fossils, the only way gaps (existing in the paleotological record) can be closed is by the theory of morphology with the useless argument that because certain characteristics of animals may be homologous, the animals must have evolved from each other."

Darwin had this to say about morphology and homologous structures:

"This is one of the most interesting departments of natural history, and may almost be said to be its very soul. What can be more curious than that the hand of a man, formed for grasping, that of a mole for digging, the leg of the horse, the paddle of the porpoise, and the wing of the bat, should all be *constructed on the same pattern*, and should include similar bones, *in the same relative positions?*" [Emphasis mine]

The Origin of Species

Pp.334-335

Remarkable! – Darwin asks the question concerning homologous structures and then is inconsistent in his conclusion. He observes and is more than curious that morphological comparisons with the bones of man and animals are constructed "on the same pattern" and "in the same relative positions" but the relativity to all this, escapes him. For many persons of Darwin's day, this curious bone construct would not only indicate a pattern but a Designer as well. Of course, Darwin was too much into his theory to be bothered by a notion such as a pattern designed by a Creator. Darwin believed the question of patterns could be explained by his theory of modification. This in fact was no explanation but simply replacing the uniform plan of a Creator with the theory of natural selection. Strange! – Darwin regarded theory as a scientific explanation.

In fact Darwin, later in this section on "morphology," mentions Professor Richard Owen who, as a creationist, held to the pattern belief:

"The hopelessness of the attempt has been expressly admitted by Owen in his most interesting work on the 'Nature of Limbs.' On the ordinary view of the independent creation of each being, we can only say that so it is; - that it has pleased the Creator to construct all the animals and plants in each great class on a uniform plan; but this is not a scientific explanation."

P.334

Darwin's *inconsistency* came with his rejection of the *innate truth* (patterns and a Creator) for the acceptance of the philosophical error of attributing complex organs and structures to the realm of *genetic chance*.

[It is not my intention to assume that the Creator and His design should be irrefutably evidenced in the natural world; rather this *innate truth* is the truth of reflective reasoning evidenced by the acceptance of the certain operation of intelligent purpose, other than the acceptance of the *unpredictable and uncertain theorem of blind chance*]

Darwin took the pattern theory and replaced it with his own theory of modification. What was the *additional inconsistency?* – It was Darwin making the charge that Owen's explanation was not scientific but his own theory, by overt necessity, was scientific. Darwin's stubborn and committed stance in evolution is witnessed today in the camp of evolutionary paleontologists who will believe anything in support of the confirmation of their theory, even the transformation of a land quadruped into a whale.

CHAPTER FOURTEEN

DARWIN'S FINCHES VISIT
GALÁPAGOS ISLANDS

In the *National Geographic*, November 2004, pages 26-27, David Quammen, refers to the "ground finches" and their diversification through isolation, as a major proof of evolution.

IN THE GALÁPAGOS ISLANDS IN 1835, DARWIN COLLECTED SOME SMALL BROWNISH BIRDS OF VARIOUS SIZES AND SHAPES IN THEIR BEAKS ..."GROUND FINCHES," MORE THAN A DOZEN NEW SPECIES, UNKNOWN TO SCIENCE (P. 26).

Comments: - The Galápagos Islands are some fifteen to twenty miles apart. Darwin observed that each island was occupied by many different species of birds with some diversification. For example, Darwin recorded thirteen species of ground finches. He saw that there were differences in the beaks, tails, and plumage of these finches – variations in nature. Yet, the finches from the various islands resembled one another. In other words, Darwin's finches were not separate species but ordinary variation within a species.

THERE WAS A SIMILAR PATTERN OF DIVERSIFICATION, DARWIN HAD NOTICED, AMONG GALÁPAGOS TORTOISES AND AMONG MOCKINGBIRDS.

Comments: - The land tortoises were represented by fourteen species on the nine largest islands. The islands take their name from these giant land tortoises.

Marsh writes:

"Here again we must keep our feet on the ground of solid facts and recognize that these changes are only occurring within the boundaries of the separate kinds. The tortoises, for example, merely develop other 'species' of tortoises … There is no real evidence that changes from one kind to another have occurred."

Evolution, Creation, and Science

Frank Lewis Marsh

Pp. 301-302

The same might be said for other animals as well – the mockingbirds, the lizards, the finches, etc. These creatures, like the tortoise, merely developed "species" after their own "kind."

WHY SHOULD REMOTE ISLANDS CONTAIN SUCH DIVERSITY? HIS [DARWIN'S] ANSWER WAS THAT ISOLATION – PLUS TIME, PLUS ADAPTATION TO LOCAL CONDITIONS – LEADS TO THE ORIGIN OF THE SPECIES. IT SEEMED MORE LOGICAL THAN ASSUMING THEY HAD BEEN CREATED AND PLACED IN THE GALÁPAGOS INDIVIDUALLY.

Comments: - David Quammen, the author of this *National Geographic* article, was not being completely candid when he wrote "It seemed more logical than assuming they had been created and placed in the Galapagos individually." These words are more than his casual after-thought. They were indicative of what Darwin, at one time, actually believed. Darwin believed in <u>static distribution</u> or <u>fixation</u>. Quammen, the author, does not wish to attribute this belief to Darwin. It is Quammen's attempt to protect Darwin's reputation as an impeccable scientist. In divinity school, Darwin had been taught that <u>species are immutable</u>. He accepted the <u>static theory of distribution</u> by an act of faith in the teachings of men and by denying his faith in the Master Instructor who offered the biblical book of Genesis for His textbook. The early scholars taught that God created each animal and set it, individually, within its particular habitat. Darwin and his instructors failed to understand that the true intent of the Genesis account did not teach the immutability of species but the exact opposite - VARIATION of species. It is hard to believe that Darwin attended a school of divinity!

Darwin, after observing species on the Galápagos Islands, in one month's time, thought it to be unreasonable to conclude that only one species should have been created for each small island. In time, his deductions, in answering the question on how the different species arose, led him away from the scholastic theory of "fixation" without variation. Although I believe that Darwin did not become an evolutionist during his voyage on the Beagle (1831-1835), I do believe that his observations at the islands correctly convinced him of variation.

David Quammen implies that Darwin's research on finches lead him to the conclusion that these birds revealed something about the origin of species. On the one hand, Quammen tells the public that isolation, plus time, plus adaptation to local conditions, led Darwin to the origin of species.

On the other hand, Jonathan Wells informs us:

"Yet the Galapagos finches had almost nothing to do with the formulation of Darwin's theory. They are not discussed in his diary of the *Beagle* voyage except for one passing reference, and they are never mentioned in *The Origin of Species.*"

Icons of Evolution Science or Myth

P.160

I wonder if Quammen bothered to research Darwin and the finches! I don't believe he did or he would have soon discovered the informative statement by Wells to be absolutely true.

I went to my library and selected *The Origin of Species* and *The Voyage of the Beagle* in order to investigate the claim of Wells. *The Voyage of the Beagle* is a renaming of Darwin's *"Journal of Researches"* (according to Peter Nichols in his book *Evolution's Captain* and page 192). Here is what I discovered:

In the *Beagle* journal, Darwin mentions a thick-billed finch picking at one end of a piece of cactus (P.337) and the extreme tameness of finches (P.344). In these brief descriptions, Darwin offers no official opinion concerning finch evolution. Darwin's journal was published in 1887, twenty-eight years after *Origin of Species*; plenty of time for Darwin to think about incorporating his evolutionary views on finches into this same journal but he did not. What was the reason for his not doing so? — he was not impressed with the evidence. Besides it may come as a surprise to most people, Darwin was not a student of finches; he had a limited and often an erroneous conception of finches; he knew little about their feeding habits and geographical distribution of these birds.

The following passage is the only statement that discusses the island finches:

"The remaining land-birds form a most singular group of finches, related to each other in the structure of their beaks, short tails, forms of body, and plumage: there are thirteen species, which Mr. Gould has divided into four sub-groups. All these species are particular to this archipelago... [On page 328, there is an illustration of four species of birds with various beak sizes but the illustration has no comments about natural selection or evolution].

The Voyage of the Beagle

Charles Darwin

P.327

Is there anything in these words that would lead us to believe that Darwin had formed a theory about the origin of species? — not one word! At this point, Quammen is far from the truth and Wells is focused on the truth when he wrote, "Yet the Galapagos finches had almost nothing to do with the formulation of Darwin's theory."

I picked up from my desk, *The Origin of Species*. Surely, if Darwin developed a theory about finch evolution, it would be recorded in this book — the bible of evolutionary philosophy. As I searched carefully through the book, not a thing was recorded about finches. The only index reference about finches was not from *The Origin of Species* but from *The Descent of Man*. Darwin mentions the coloration of both sexes in *British* finches and that is the extent of his comments on finches.

Wells is correct. Darwin did not use the Galapagos finches as supreme examples of evolution as Quammen believes. But research is not a matter of who is right and who is wrong. It is a matter of honest investigation on the part of individuals who attempt to arrive at the truth or get as near to the truth as they possibly can. Apparently, Wells arrived at truth because of his intensive investigation.

In our calling the species of birds "Darwin's finches," Darwin never, during his life, heard of the expression. In 1947, David Lack wrote a book bearing the same title and so we use the title anachronistically when we link it to Darwin.

Wells gives further information:

"Although they were first called 'Darwin's finches' by Percy Lowe in 1936, it was ornithologist David Lack who popularized the name a decade later. Lack's 1947 book, *Darwin's Finches*, summarized the evidence correlating variations in finch beaks with different food sources and argues that the beaks were adaptations caused by natural selection. In other words, it was Lack more than Darwin who imputed evolutionary significance to the Galapagos finches. Ironically, it was also Lack who did more than anyone else to popularize the myth that the finches had been instrumental in shaping Darwin's thinking."

Icons of Evolution Science or Myth?

Pp.162-163

[Quammen fell for this myth and he hoodwinked the public into accepting the same myth] However, there is one important fact of which we should be aware. The finches are being used by modern researchers to get across Darwin's theory of natural selection and how it is a possible mechanistic force for the origin of species. This is the reason for Quammen's elaboration on the same issue. There are evolutionists who think that the clues for the doctrine of transmutation are wrapped up in the shapes of finch bills. Rather than seeing variation as the power of God manifested in the finches' ability to adapt to most circumstances in nature's ecology, by using their bills as tools; their minds dismiss God's power and turn to natural selection as the mechanical force behind evolution.

Today, they reason, a change in the shape of a bird's bill could lead tomorrow (geologically speaking) to a different animal altogether. Their field notebooks direct them to make unwarranted decisions but modern scientists, who research finches, know nothing about the genetics of finch beaks. There is no direct evidence for genetically linking finch bills with evolutionary ancestors of the past or for believing that variation in bills, plus time, will be included in with the total mutational changes necessary to change or evolve a finch into some other life form.

The chromosome studies show no differences among the Galapagos finches. Finches are birds and although there are great varieties of finches they will never be anything but the bird-kind. The finches show speciation because they have become reproductively isolated from their parent population but they will be forever limited in their variation.

Speciation will always demonstrate limited diversity on the horizontal plane but nothing will ever be added to the gene pool that would shape finches into another kind of animal. The pattern of finch beaks, correlated with food sources, has become a major source of study for evolutionists attempting to find in natural selection the explanation for origin of species among Darwin's finches. But such studies will never be able to show scientists anything other than the demonstration of variation without transmutation. The study of a million finch beaks will never transmute the finch into any animal that would not be recognizable as a bird.

What a tragedy for Darwin! As a young man, his divinity school granted to him the unfortunate legacy of a narrow concept of creation without the capacity for each created species to vary. Apparently Darwin and his instructors were ignorant of the Bible record. The Bible did not teach that everything created was set statically in its present environment; that God did not allow for variation of plants and animals; that animals could not breed abundantly and produce variants from a common stock – the "kinds."

Studies were carried out in the 1980s that indicated at least half of known species of finches on the Galapagos are known to hybridize which, in simple terms, is crossbreed. That is, two animals of different species or plants of different variety being able to mix and produce offspring called hybrids. Hybridization can designate both narrow crosses and wide crosses. Evolutionists love to call this type of breeding, evolution in the making. However, "the most that hybridization can do in the matter of change is to give rise to another variation *within some already existing kind.* This is *not* evolution in the sense that evolutionists use it in their theory." (Marsh in *Evolution, Creation, and Science,* P.139)

Quammen speaks about *isolation* of the finches and how natural selection, plus time, creates new species. But isolation viewed as leading to the stages of evolutionary development, is pure philosophy. There is no scientific evidence available which proves that the physiological isolation of finches on the Galapagos Islands accomplishes the creation of new kinds. The only thing accomplished by such a process is the mere increased complexity of the kind.

Returning to Darwin's academic schooling, it can be said again, what a tragedy! Darwin, in his restricted view of nature, rejected the biblical account of creation; in his answer to diversity, refused to see Galápagos as one aspect of the Creators plan for scattering the animals and, under isolation, having them diversify with limitation; in his Galápagos observations, missed that animals may constantly alter their general appearance while at the same time having this alteration and variation limited.

Back in the 70s, I attended Andrews University in the state of Michigan. I had the pleasant experience of attending a class in creationism taught by the paleontologist, Harold G. Coffin. James Perloff quotes Coffin in his outstanding book – *Tornado in a Junkyard*, P.47 -"We must admit that Darwin did see different variations from one island to another in the Galápagos. And he did see evidences that made it necessary for him to discard a belief that living things did not change. But they were relatively minor and he had no compelling evidence that forced him to believe in limitless transformation. Darwin made a common mistake – that of 'either – or.' Either species were fixed or unlimited change occurred. But the truth lies between the two extremes," *Origin By Design*, P. 344.Unfortunately, Darwin came to think of the Galapagos Islands as laboratories with animal species serving as experiments in evolution with unlimited variation.

In a final word concerning Darwin's finches:

"And God created … every winged bird according to its *kind*. And God saw that it was good … God blessed them and said … let the birds increase on the earth."

Genesis 1:21

The Bible teaches that change (variation) is permissible but definite limits are established for it at the boundary line of the kind. In other words no matter how many varieties, birds will always be birds. Darwin was a breeder of 90 fancy pigeons but he always ended up with pigeons. Darwin was an observer of finches but no matter how much their bill differed among the variations, the finches would always be known as birds and nothing else.

Darwin *was wrong* in teaching evolution; man is wrong in calling the finches, Darwin's finches. The finches belong to God – they are God's finches and God *was right* when He said:

"Do you not know?

> Have you not heard?

The LORD is the everlasting God,

> The *Creator* of the ends of the earth."

> Isaiah 40:28

Evolutionism is a transitory and fleeting doctrine that is soon to pass on;

Creationism is a lasting belief centered in the eternal and undying God.

CHAPTER FIFTEEN

BIRD FLIGHT, MAN'S INTELLIGENCE
AND THE VERTEBRATE EYE

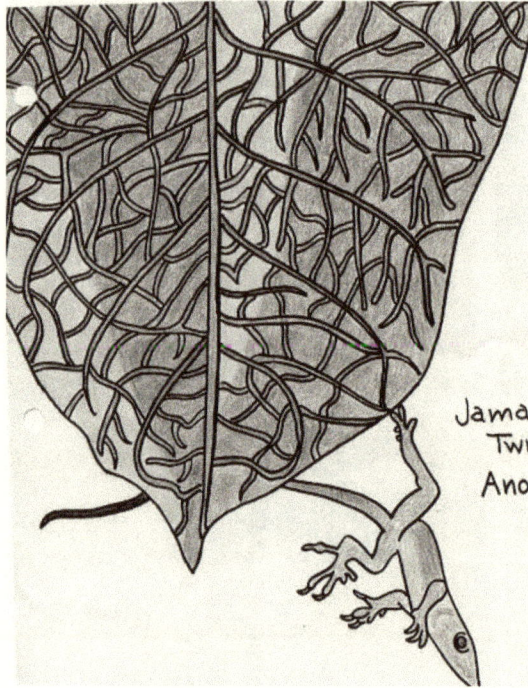

Jamaican Twig Anole

In the *National Geographic*, November 2004, pages 28-29, David Quammen cites the Jamaican twig anole and how it strikingly resembles the Puerto Rican twig anole and the Hispaniola twig anole. These animals developed striking similarities resulting from their adaptations to similar ecological conditions. Evolutionists call this convergent evolution.

Quammen's vision has been darkly clouded by that unstable theory that deems animals, with less than complex morphologies evolve easily and arise again and again. Such thinking has a low opinion of simple forms (as though simple forms do not involve all the complexities that go into the making of any life form) and a high opinion of adaptability in the realm of natural selection.

In *Webster's Family Encyclopedia*, P.647, convergence (or convergent evolution) is defined: "The development in unrelated animals of similarities resulting from adaptations to the *same way of life*. Thus whales (mammals) and fish have evolved similar features independently, associated with their aquatic habitat."

[Emphasis mine]

This definition states that organisms from different backgrounds can develop similarities because they live in a similar environment with the same selective forces influencing their heredity. However, this definition comes up short in describing the full meaning of convergence.

There are other types of convergent evolution mentioned by Harold G. Coffin in *Accident or Design*: Species can be from *two separate environments*, yet are similar. Examples: - Hermit crabs of Washington State are matched with hermit crabs along the coast of Chile; two separate environments, yet the hermit crabs are remarkably similar. This is partially explained by the fact that although two separate environments are involved, the *physical factors in nature are much the same.*

There is yet another type of convergent evolution. Similarities appear in unrelated species (with no phylogenitic relationships) from two separate environments. This is harder to explain for the theory of evolution because the *physical factors in nature are not the same*, such as temperature, oxygen, food supply, etc. Examples:-

* The vertebrate and the cephalopod both have the same type of eye. These structural similarities must have arisen independently.

* Fish, toothless anteaters and birds all have gizzards. The only way this structural similarity can be accounted for in these different animals, is to conclude that gizzards must have arisen independently.

* Termites (Isoptera) and bees, wasps, and ants (Hymenoptera) are separated by so large a taxonomic unit that there is no possible way of their maintaining common relationships. Yet, both these taxa have workers, soldiers, kings, and queens. Again, the theory of evolution must engender *faith* in order for evolution to explain these similarities by converging evolution. The chance of these two units developing a social order with similar morphological types is so remote that it must be embraced by *faith* alone.

(Coffin, Pp.410-412)

* The Ichthyosaurus is an extinct "fish reptile" that according to evolutionists, existed in the Triassic seas 250 million years ago. The first

Ichthyosaurus remains were discovered in England at the end of the nineteenth century. This fossil is similar to the living shark. The fishlike features, in both these reptilian forms, is the product (so evolutionists claim) of convergent evolution. This claim again, calls faith into play because, structurally, these "fish-reptiles" were far removed from any other reptilian order.

(See *A Search for Meaning in Nature*, Richard M. Ritland, P.289)

The late evolutionist Stephen Gould writes:

"But convergence, however stunning in general adaptive features of basic form and function, can never be intricately precise in hundreds of detailed and highly particular parts – because converging lines begin from such different antecedents and must craft similarities from disparate starting points. Thus, ichthyosaur paddles may be dead ringers for fish fins in external form. But they are built of finger bones from a terrestrial past; and the eyes of squid and vertebrates, though so similar in final form, follow markedly different embryological routes in their construction."

Dinosaur in a Haystack

P.118

In other words, certain animal structures may appear to be homologous but, as Gould has brought out, the concept of homology cannot be traced back into embryology. Ichthyosaur paddles and fish fins, eyes of squid and vertebrates are arrived at by different routes and are not really homologous in the Darwinian sense. [That is, such homology is in no way suggestive of true relationship of inheritance from a common ancestor]

Michael Denton in *Evolution: a Theory in Crisis* has written:

"It appears then that Darwin's usage of the term 'homology', which he defines in the *Origin* as that relationship between parts which results from their development from corresponding embryonic 'parts', is, as DeBeer emphasizes, just what homology is not. The evolutionary basis of homology is perhaps even more severely damaged by the discovery that apparently homologous structures are specified by quite different genes in the different species. The effects of genes on development are often surprisingly diverse."

P.149

Similarities in structures (comparative anatomy) cannot trace ancestors through embryogenesis or genetics. The study of Darwinian homology arrives at a blank wall when using it as a "proof" of inheritance from a common ancestor – one more nail in the coffin of evolution. More logical is the assumption that these similarities were all designed and created by intelligence than it is to believe that similarities arose by chance with the involvement of countless, random genetic errors.

Stephen Gould wants his say in this matter but goes *beyond circumstantial reasoning* to the point wherein his opinion becomes *ecclesiastical* (Darwin never attempted to "prove" evolution. He sought to establish his theory by *circumstantial reasoning*). Gould's *faith* in natural selection is noteworthy. He elevates natural selection to omnipotence and claims that it has power "to craft optimal structures." Three times he mentions the word *power*. Gould is guilty of doing what he so often accused creationists of doing – turning science into a religion. He has natural selection take the place of God:

"If adaptation and natural selection wield such *unimpeded power* over the fate of each evolutionary sequence, why search for deeper commonalities in lineages long separate? Arthropods and vertebrates do share several features of functional design. But those similarities only reflect the *power of natural selection* to craft optimal structures independently in a world of limited biomechanical solutions to common functional problems –an evolutionary phenomenon called convergence... ...natural selection ... ... holds the capacity to build such intricate convergences as independent illustrations of its *predominate power*." [Emphases, mine]

Leonardo's Mountain of Clams and the Diet of Worms

P.32

Gould believes that natural selection is capable of doing some marvelous things. If two different kinds of animals live in separate ecologies, then natural selection not only has power to shape a structure to meet a certain animal's need but has the power to shape the same structure that already exists as a characteristic in some other animal, when the ecology requires such a similar adaptation.

To explain such parallelism, through the phenomenon of converging evolution, calls for faith; to believe that highly adaptive forms can arise again and again by evolution, calls for a more potent brand of faith; to believe that similar structures can arise in diverse animals from chance mutations and by natural selection, is to believe in the impossible.

One more quote from Gould before considering both quotes together and in context:

"... from the phenomenon known as 'convergence' ... Flight has evolved separately in insects, birds, pterosaurs (flying reptiles), and bats. Aerodynamic principles do not change, but morphologies differ widely (birds use feathers; bats and pterosaurs employ a membrane, but bats stretch it between several fingers, pterosaurs only from one) ... Since adaptive themes are limited and animals so diverse, convergence of different evolutionary lineages to the same general solution (but not to detailed repetition) are common. *Highly adaptive forms that are easy to evolve arise again and again.* More complex morphologies without such adaptive necessity offer little or no prospect for repetition. Conscious intelligence has evolved only once on earth ... Perhaps, in another form on another world, intelligence would be as easy to evolve as flight on ours ... the strong 'classical' argument for convergence – that the eyes, although so similar in design and operation, develop embryologically in fundamentally different ways (squid eyes form from skin precursors, while vertebrate eyes, the lens excepted, develop from the brain)." [Emphases mine]

The Flamingo's Smile Reflections in Natural History

Pp.411-12

Comments: -Two long quotes so the readers can understand that natural selection does in fact take the place of God - Three majestic, phenomenal aspects of creation are attributed to the "power of natural selection" and has it "craft optimal structures" instead of God. Natural Selection is personified and through its anthropomorphism begins to craft or fashion: - Flight, Intelligence, and Vision.

Only God has the power to perform these creative acts:

The prophet, Jeremiah, wrote:

"But God made the earth by *his power,* he founded the world by his wisdom and stretched out the heavens by his understanding ... for he is the Maker of all things ... (10:12, 16)

"This is what the LORD Almighty, the God of Israel, says: 'Tell this to your masters: With *my great power* and outstretched arm I made the earth and its people and the animals that are on it ...'" (27:4-5)

The apostle, Paul, penned:

"For since the creation of the world God's invisible qualities – *his eternal power* and divine nature – have been clearly seen, being understood from what has been made, so that men are without excuse ... They exchanged the truth of God for a lie, and worshiped and served created things rather than the Creator – who is forever praised. Amen." (Romans 1:20, 25)

The beloved, John, exclaimed:

"After this I heard what sounded like the roar of a great multitude in heaven shouting: 'Hallelujah! Salvation and glory and *power belong to our God,* for true and just are his judgments'". (Revelation 19:1)

Let us speak to those issues which Stephen Gould attributes to the power of natural selection – Flight, Intelligence, and Vision:

(1) Bird Flight – Bird wings will serve as one of the examples of flight that has developed (According to Gould) through the *phenomenon of convergence. This is such an "easy" phenomenon for Gould! – For him, natural selection* takes lizard scales and makes those into sophisticated feathers; eventually having those feathers transformed into air-worthy wings for an evolved lizard, which has become a bird (Gish); in one geological breath, regulate the bird instinctively for aerodynamic skills.

We need to turn to the advice of Job:

"But ask the animals, and they will teach you, or the birds of the air, and they will tell you..." (12:7)

What can the birds tell you? – Their anatomical make-up and adaptability proclaim the marvelous power of a loving Creator.

Bird anatomy – the bones are hollow, a perfect characteristic for flight, especially for a large bird; the blood pressure is high and rapid energy can be delivered to the breast muscles; the respiratory system has air sacs which allow continuous delivery of air (The birds have a different respiratory system from that of reptiles. Because of how they function, there can be no intermediate link between reptile lungs and bird lungs. Birds did not evolve from reptiles).

Flight feathers have veins, barbs, flanges and tiny hooks that work together as a zipper. When it becomes unzipped, it can be rezipped for primping.

Evolutionists call this type of anatomy, chance and convergence; others call it, the miracle of design.

Soaring flight – a good example is the stork. In Jeremiah 8:7 "Even the stork in the sky knows her appointed seasons ..."

The prophet spoke of the stork's migration. The stork rides air currents throughout its annual journeys from the Baltic to South America. The bird flies into a rising twirl and it must linger in the column of rising air. The bird breaks away from the top of the spiral into a long and graceful gliding motion. All the flight skills of gliding, soaring, flapping, and hovering involve the sophisticated use of aerodynamics.

Evolutionists call this ability to fly, evolution in convergence; others call it, the miracle of design and creation.

Bird migration – migration is a predictable event. Here, in southern California, we have the cliff swallows of San Juan Capistrano Mission. How does a bird know when to migrate? When the day reaches a certain length, the bird's hormones reach high level and trigger the urge to migrate; when blown off course by a storm, the bird has a "homing ability" and will still find its way over unfamiliar territory; the young birds, with no training have a "genetic map"; at night, the bird in flight, uses star patterns; for day flight, the bird is guided by mountain ranges, rivers, shore lines, etc.

Evolutionists call this remarkable migrational trait, evolution and convergence; others call it, the miracle of design.

In Job 12:9, 10: "Which of all these does not know that the hand of the Lord has done this? In his hand are the life of every creature and the breath of all mankind."

Evolutionists call the anatomy of the bird, chance; the bird's ability to fly, selection; the migrational trait of the bird, evolution; others call all these "crafted optimal structures and functional designs" testimonies to the Master Designer.

(2) Intelligence – According to Gould, wings were "easy" to evolve but "conscious intelligence" evolved only once on earth with "little or no prospect for repetition." Conscious intelligence is a complex morphology but Gould could not leave well enough alone. He had to say, "Perhaps, in another form on another world, intelligence would be as easy to evolve as flight on ours." Faith comes easy to the evolutionist.

God gave to human kind intelligence – the ability to remember and utilize past experience in order to cope with current situations; to think and reason with unprecedented power; to conceive of new thoughts and ideas based on previous observations. Intelligence clearly differentiates humans from all other animals. This is why man, contrary to the thinking of evolutionists, should be placed in a different order separate from the animals.

Think about it! – Our brain works far better than any man made computer and can store a lifetime of information. No wonder that evolutionists have to admit conscience intelligence evolved only once on earth and with "little or no prospect for repetition." Our brain contains around 100,000,000,000,000 connections with nerve cells. How can men reason that this complicated organ, with its amazing interdependent parts, could have its origin by random evolution?

Evolutionists call it, complex morphology and evolutionary convergence; others call it, created in the image of God.

(3)Vision – Gould contends that squid eyes and vertebrate eyes are the strong "classical" argument for convergence; that the eyes of both animals are similar in design and operation; that they, however, are fundamentally different in their embryological development.

I read over twenty pages in Gould's *The Structure of Evolutionary Theory* hoping to learn something (from the evolutionist's point of view) about the origin of the human eye. I carefully scrutinized each paragraph but in spite of all that technical material, I came away empty. With all that man knows about how the eye works, Gould could not explicate from any scientific writings, including Darwin's, how the eye came into being. As much as he supported the Darwinian doctrine of natural selection, Darwin offered Gould no clue in the *Origin of Species* under "Organs of Extreme Perfection and Complication."

Darwin (with a crude sketch) attempted to convince the public of the complexity of the human eye by bringing it through the various stages of evolution – not in human development but in animal development - and in the final stage of the sketch, pictured the amazing human eye standing alone in all its glory. In other words, Darwin attempted to show the development of the eye in its most simple stages in the less complex of animal forms and in the final stage of the eye's evolution, show the human eye. This illustration was not only "slight of hand" on the part of Darwin but just plain deception - deception of the highest order. The imagery was simple in form but broad in its purpose   to hoodwink the gullible and the unwary public. Darwin's sketch was superceded only by Haekel's deceptive drawings of the human embryo.

It is remarkable that with the advancement of science and all the light man has gained in his knowledge concerning the human eye, some advocates for evolution, nevertheless, have accepted Darwin's vain and superficial way of explaining the organ of sight. How Darwin beguiled people into believing the eye would take less than the power and creatorship of a Divine Architect to fashion, *is still a mystery.* For a scientist to have a following of disciples who actually believed that the evolutionary pathway of the organ of sight began in some creatures head as a simple light-sensitive spot and grew into the breath-taking, elegant, and sophisticated television camera-eye of man, *is a greater mystery.* With all of Darwin's and Gould's ramblings, the eye really cannot be explained to anyone's satisfaction.

Michael J. Behe makes this comment concerning Darwin:

"He did not even try to explain where his starting point – the relatively simple light-sensitive spot – came from. On the contrary, Darwin dismissed the question of the eye's ultimate origin: 'How a nerve comes to be sensitive to light hardly concerns us more than how life itself originated.' He had an excellent reason for declining the question: it was completely beyond nineteenth-century science. How the eye works – that is, what happens when a photon of light first hits the retina – simply could not be answered at this time."

Darwin's Black Box

P.18

Darwin, at this point in life, must have realized the impossibilities of his own theory. He wrote in *Origins*, P.135, sixth edition, "If it could be demonstrated that any complex organ existed, which could not possibly have been formed by numerous, successive, slight modifications, my theory would absolutely break down. But I can find out no such case."

Professor Michael Behe, the biochemist, has pointed to *many cases.* The readers might refer to Behe's *Darwin's Black Box*, Pp.15-22, where the evidence is overwhelming that the eye could not be produced from the simple to the complex. The eye must not only be explained anatomically but it must be explained as well from the standpoint of biochemistry. There is no way to account for the complexity of the human eye through Darwin's evolutionary pathway of incipient steps. The eye is an irreducible organ; it must be set in place all at once and ready to go; and, if not, it would be totally useless in each stage of its evolutionary sequence.

Evolutionists call the eye, the organ of extreme perfection. Creationists also call it that but they know where this perfect design came from and Who designed it.

Darwin wrote:

"…The belief that an organ so perfect as the eye could have formed by natural selection, is enough to stagger anyone."

P.181

Even evolutionists reel and stagger at this prospective of the perfection of the eye but their faith in natural selection is always strong enough to sober them up and bring them back to the natural world and to their studies containing the books that adhere to the unnatural teaching of evolution; to the unrealistic and vain philosophy that attempts to explain the human eye and other life's origins, without the power of God.

Darwin had a full idea of the ingenuity of the optical system through his knowledge of the camera and telescope but he did not realize what we now know – the eye is a far more sophisticated instrument than it appeared a hundred years ago. Michael Denton claims that "today it would be more accurate to think of a Television camera if we are looking for an analogy to the eye." *Evolution: a Theory in Crisis*, P.333

In the case of flight, intelligence, and vision, Darwin's imagination conquered his reason. Natural selection and convergence cannot possibly account for these wonderful and astonishing feats of creation. Apparently, imagination and not facts has captured *all evolutionists* from the time of Darwin and onward to this present day.

CHAPTER SIXTEEN

THE FRUIT FLY – ONCE A DROSOPHILA
ALWAYS A DROSOPHILA

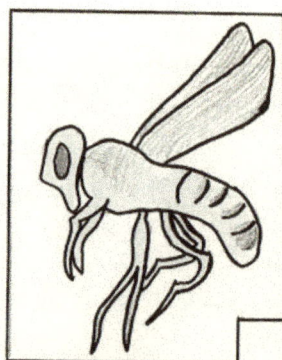

Drosophila melanogaster

176

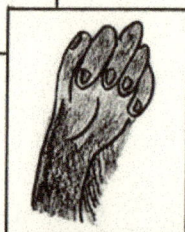

Chimpanzee Hand

RBP

In the *National Geographic*, November 2004, page 32, David Quammen, makes the following comment:

GREGOR MENDEL, AN AUSTRIAN MONK, DISCOVERED THE FUNDAMENTALS OF GENETICS IN DARWINS TIME, BUT HIS IDEAS, PUBLISHED IN AN OBSCURE JOURNAL, WERE IGNORED. LATER BIOLOGISTS MERGED EVOLUTIONARY THEORY WITH GENETICS.

Comments: - What a great temptation to run with the ball on this one! But, remarks will be confined to a short comment on Quammen's misleading reflections.

Quammen takes great pains to cover up the truth that evolutionists were themselves mostly responsible for the suppression of Mendel's great discovery in genetics. I have the following notes sketched around the picture of Mendel on page 32 of the *National Geographic*. These short notes express the real truth of what happened historically in the days of Darwin and evolutionary biologists:

"It may have been true that Mendel's ideas were ignored. Darwin should know – Mendel had sent Darwin a copy of his paper and Darwin completely ignored Mendel's work. Darwin had a copy of Mendel's document in his library, which he never read. A recent examination showed that its joined pages were left uncut. And evolutionists should know – Mendel's work was published immediately after 'The Origin of Species' and his work was completely ignored by evolutionists for fifty years. 'Later biologists merged evolutionary theory with genetics ...' Yes, but what Quammen left out was, the theory became NEO-DARWINISM – A theory that included Mendel's work."

And how much has Neo-Darwinism contributed to the cause of evolution? The answer by many authorities of the evolutionary persuasion is, "not much" (And this is in no way a reflection on Mendel's scientific findings).

For one example that could be quoted among many, Christian Schwabe "On the Validity of Molecular Evolution," *Trends in Biochemical Sciences* (July 1986) claims:

"The neo-darwinian hypothesis, in fact, allows one to interpret simple sequence differences such as to represent complex processes, namely gene duplication, mutations, deletions and insertions, without offering the slightest possibility of proof, either in practice or in principle."

The contents of the above quotation can hardly be considered to be a contribution!

In the *National Geographic*, November 2004, page 32, David Quammen refers to the vinegar fly as a supporting proof for evolution.

Comments: - The small vinegar fly, Drosophila melanogaster, "is the geneticist's godsend; the great laboratory stalwart; harboring four pairs of chromosomes per diploid cell" (Stephen Gould comment).This means the geneticist is enabled to conduct many experiments within a short time period because of the fly's ability to rapidly mutate. Therefore, many genetic mutations of the fruit fly can be produced and studied in laboratories.

Through this process of mutation, superficial characters may be noted in the fruit fly. Certain changes may be observed in the fly's appearance such as the coloration of the external and internal parts, the length, diameter, and shape of the bristles, etc., etc. However, all these changes are not major. Hundreds of mutations have been recognized in the Drosophila and each mutated form is still, nevertheless, a fruit fly.

What do these experiments prove as genetic changes are observed in the fruit fly? In nature, the mechanisms for variation are slow and uncertain. That is, mutational changes can be fatal or cause deteriorations but even as gene and chromosomal changes, normally working at a slow rate, can be speeded up through experimentation in laboratories, nevertheless, the transformation of the Drosophila fly into another type of species or animal has never been observed.

The fruit flies prove variation on the micro (small) level but the mechanisms that produce such varieties cannot and never will provide a satisfactory explanation of changes that might lead to new families or higher taxa. The fruit fly, try as it will, by modeling her different garb and reaching for that great change (mega evolution) will always and predictably fall short. On page 32 of the *National Geographic* I have these brief notes at the bottom of the page:

"Once a fruit fly, always a fruit fly; proving that variations has limitations; with all his 1000 + mutations, Drosophila melanogaster has never achieved anything other than a vinegar fly."

Thomas Hunt Morgan, beginning in 1900, bred tens of thousand of these fruit flies in milk bottles in the laboratory. He was determined to study the transmission of hereditary characteristics. In 1926, Morgan wrote his *Theory of the Gene*. Out of his work arose Neo-Darwinism which is Darwinism interpreted in terms of Mendelism. Nevertheless, the cause for variation was still not answered.

Once again we come to that question of *faith*. William Bateson, a British biologist, founder of the science of genetics, president of the British Association for the Advancement of Science, 1914, remarked:

"We cannot see how differentiation of species came about. Variation of many kinds, often considerable, we daily witness but no origin of species. Meanwhile, through our faith in evolution stands unshaken, we have no acceptable account for the origin of species" (Bateson's statement is quoted by Harold W. Clark in his book "The Battle over Genesis." P. 118).

Evolutionary scientists can study the fruit fly for the next hundreds of years and from it they will never see another kind of animal develop. They might, by wishful thinking, guarantee us that they are witnessing evolution in action. The need for faith in the reasoning of the evolutionist has been evidenced in the various sections of this book. It takes *faith* in surmising fruit fly will become something different as the years mount up; *faith* to postulate a giraffe has an evolutionary ancestor when the fossil record proves otherwise; *faith* to give credence to the Darwinian tree of life when the twigs and limbs do not really connect with the trunk; *faith* to assume the laws that state "life only from life" and "like begets like" are not in keeping with the precepts of biology; *faith* in maintaining the belief that a flying fish can be modified into a perfectly winged animal; *faith* in trusting a whale has an antelope for an ancestor; etc., etc. In spite of what David Quammen writes, "No one needs to, and no one should, accept evolution merely as a matter of faith," we know better. With the facts of evolution falling short, how else should evolution be accepted if not through faith?

Of the many mutations discovered in the fruit fly, some cause the normally two-winged fruit fly to develop a second pair of wings. This feature has been capitalized upon in textbooks and even in public meetings as a representation of evolution. These mutant flies are not a product of nature but are the result of breeders in a laboratory who perform this mutational experiment under strict and rigid control.

Jonathan Wells brings up the fact that Peter Raven and George Johnson's 1999 textbook *Biology* features a photo of a four-winged fruit fly and attempts to make a case for evolution by stating that the re-arrangement of existing genes provides the raw materials for evolution. The authors do not emphatically state that the fruit fly shows us evolution in action but they certainly imply that genetic mutations are the origin of new variations. Wells, in an excellent run down on the flaw and error of this implication, points out the following facts (Here is a general summation of his findings):

"Three separate mutations had to be artificially combined in one fly to produce a second set of normal-looking wings (This combination is not likely to occur in nature).

"Even more seriously, the textbook fails to point out that the second pair of wings is nonfunctional [Biologists have known for over five decades that the extra set of wings of *bithorax* mutants lack flight muscles]

"The hapless insect is thus disabled. And the disability increases with the size of the mutant appendages.

"Because of this, four-winged males have difficulty mating, and unless the line is carefully maintained in a laboratory it quickly dies out."

Icons Of Evolution Science Or Myth?

P.186

Four-winged fruit flies do not provide the raw materials for evolution. Even neo-Darwinists acknowledge the fact that these mutational forms are "hopeless" freaks and monsters of the laboratory. These mutants cannot find a mate and are so unbalanced that they would not stand the slightest chance of maintaining their place in nature. They would quickly be eliminated through natural selection. Wells adds that some evolutionists give "the impression that the extra wings represent a *gain* of structures. But four-winged fruit-flies have actually *lost* structures which they need for flying … … Although pictures of four-winged fruit flies give the impression that mutations have added something new, the exact opposite is closer to the truth." [Ibid. Pp.186-187]

Phillip E. Johnson is a graduate of Harvard and the University of Chicago. He has taught law for over twenty years at the University of California, Berkeley and has written a series of books against the theory of evolution. His best work, no doubt, is *Darwin on Trial*. In it he shows that the theory of evolution is based not on fact but on faith – faith in philosophical naturalism. Johnson discussed evolution for several hours in London in 1988 with Colin Patterson – an evolutionist. Patterson said he did not retract any of the skeptical statements he has made in the past. He made this admission back in November 5, 1981, at the American Museum of Natural History, New York City:

"For over 20 years I thought I was working on evolution … [But] there was not one thing I knew about it … So for the last few weeks I've tried putting a simple question to various people and groups of people. Question is: 'Can you tell me anything you know about evolution, any one thing, any one thing that is true?' I tried that question on the geology staff at the Field Museum of Natural History and the only answer I got was silence. I tried it on the members of the Evolutionary Morphology Seminar in the University of Chicago, a very prestigious body of evolutionists, and all I got there was silence for a long time and eventually one person said, 'Yes, I do know one thing – it ought not to be taught in high school' … *During the past few years … you have experienced a shift from evolution as knowledge to evolution as faith ... Evolution not only conveys no knowledge, but seems somehow to convey anti-knowledge."* [Emphases mine]

(The Patterson quote is from the book, *The Farce of Evolution*, written by Hank Hanegraaff, P.44).

The candor and frank admission of Colin Patterson, for the creationist is understandable. What is not discernible are Patterson's closing words to Phillip E. Johnson during their conversation of 1988. In spite of Patterson unveiling his true convictions about the theory of evolution, he still tenaciously and by *faith* held on to its teachings by accepting evolution as the only conceivable explanation for certain features of the natural world. How upsetting! The one explanation that Patterson acknowledged, calls for faith in the natural word – with no mention of God.

David Quammen wrote in this *National Geographic* article, "Evolution is a beautiful concept." I suppose Quammen would have a difficult time selling this concept to the public especially should the public know about the above statement of Patterson's admission.

In the end, how can a product be sold as a beautiful concept when the same product teaches at its best, "no knowledge" and at its worst, "anti-knowledge"? And, on top of all this, Quammen must sell his product on the basis of your faith, with Quammen's warning that faith is not the proper bargaining price.

CHAPTER SEVENTEEN

HOW EVOLUTION TOUCHES YOU –
NOT

In the *National Geographic*, November 2004, page 35, David Quammen makes this statement:

PETER KIBISOV, A FORMER CONVICT IN RUSSIA, CARRIES TWO ENDURING REMNANTS FROM HIS PRISON TIME: A CRUCIFIXION TATTOO AND DRUG-RESISTANT TB. HE HOPES GOD WILL HELP HIM, BUT EVOLUTION-BASED SCIENCE IS WHAT GUIDES THE SEARCH FOR AN EARTHLY CURE.

Comments: - Quammen comes to this atrocious conclusion vis-à-vis Peter Kibisov's situation! Forget about your crucifix! – Christ is no Savior; forget about God! – He can't help anybody; remember your evolution-based science as a guide for your earthly cure! But hold on, Peter Kibisov! David Quammen has just informed us that the bacteria and viruses evolve (?) so quickly, they have genetic resistance to the drug for tuberculosis.

BACTERIA AND VIRUSES EVOLVE TOO. INFECTIOUS

AGENTS SUCH AS MYCOBACTERIUM TUBERCULOSIS,

THE BACTERIUM THAT CAUSES TUBERCULOSIS, ADAPT

QUICKLY AND AQUIRE GENETIC RESISTANCE TO DRUGS

Quammen claims that bacteria and viruses evolve too. Viruses have been a confusing group. Are they living organisms? Many virologists feel that they are not actually living bodies. They cannot function and reproduce apart from living cells or cell substance. Others feel that they are some kind of degeneration of true cells.

Bacteria will now be mentioned for we know that bacteria are living entities. Quammen writes that bacterium evolves and quickly acquires "genetic resistance to drugs." Quammen uses this example of bacterium resisting antibiotics. Therefore, is a new strain of stronger bacterium created by the mutational process?

James Perloff in *Tornado in a Junkyard*, P.27, remarks:

"Mutations cause a structural defect in ribosomes – the cellular constituents that antibiotics like streptomycin attach to. Since the antibiotic doesn't connect with the misshapen ribosome, the bacterium is resistant." Perloff asks the question, "But is that evolution?" Perloff finds an answer to his question in the words of Lee Spetner:

"We see then that the mutation reduces the specificity of the ribosome protein and that means a loss of genetic information…. Rather than saying the bacterium gained resistance to the antibiotic, it is more correct to say that it lost its sensitivity to it" (*Not by Chance: Shattering the Modern Theory of Evolution* by Lee Spetner, Pp.131, 138).

Perloff had his question answered – The bacterium lost information. Therefore, this is not evolution as Quammen claims. The bacterium *did not evolve* into a more efficient mechanism of resistance. The mutational theory, harbored by Quammen, falls short of his claim for bacterium evolution.

Before getting into the question of the natural immune system that the human body has to ward off disease, let us further address the issue of bacteria. Rather than bacteria serving to substantiate the theory of evolution, bacterial characteristics afford much evidence to negate evolutionary postulates.

For example, bacteria challenge the evolutionary belief that organisms evolve through a sequence of small evolutionary steps. The bacterial flagellum defies the concept of gradualist explanations because of its complex system. The flagellum, a tiny microscopic hair, is the only structure in the entire living kingdom which exhibits a true rotary motion. Howard Berg describes the flagellum as a rotary motor. The helical filaments which compromise the bacterial flagellum rotate rapidly like propellers and are driven by a reversible motor at their base.

Michael Denton in *Evolution: a Theory in Crisis*, P.225, writes:

"The bacterial flagellum and the rotary motor which drives it are not led up to gradually through a series of intermediate structures and, as is so often the case, it is very hard to envisage a hypothetical evolutionary sequence of simpler rotors through which it might have evolved gradually."

Molecular biology has demonstrated that the tiniest bacterial cells are exceedingly complex designs. In fact, the basic design is the same all the way to mammals and is far more complicated than any machine produced by man.

I remember university days when a professor asked my class: "How many of you think there is such a thing as a simple cell?" I thought of my high school biology text book furnishing a picture of a simple cell with the usual six labeled parts: cell membrane, plasma membrane, ribosome, nuclear envelope, nucleus, and nucleolus. This professor of paleontology held an unrolled chart at eye level, about 5 1/2 feet from the floor. He let the chart unroll to the floor and said, "This, gentlemen, is a simple cell!" It is no exaggeration for me to tell you that my eyes bulged out and my mouth dropped open as I beheld the contents of a simple cell. Throughout my life and to this present hour, I no longer believe that a simple cell is simple; its complexities were breath-taking. Richard Dawkins admits, "There is enough information capacity in a single human cell to store the Encyclopedia Britannica, all 30 volumes of it, three or four times over" (*The Blind Watchmaker*, 1987).

My experience taught me that when evolutionists write about a "primitive" cell, there is no such thing. The tiniest bacterial cells are miniature factories "containing thousands of exquisitely designed pieces of intricate molecular machinery made up altogether of one hundred thousand million atoms." [Ibid. P.250]

Hickman, in "Integrated Principles of Zoology", P. 43, notes:

"Cells are the fabric of life. Even the most primitive cells are enormously complex structures that form the basic units of all living matter. All tissues and organs are composed of cells. In a human an estimated 60 trillion cells interact, each performing its specialized role in an organized community."

Creationists believe that harmful bacteria have arisen, after the Fall of Man, through degeneration. Henry Zuill, Ph.D., has stated:

"When humans sinned, it looks like they opened the 'flood gates' of natural disaster … …Degenerate environments, including degenerate occupants, stressed other degenerate inhabitants, which in turn reciprocated the stress, and so on. Some species disappeared, others adapted to new and trying conditions. Predators and parasites developed as they and ecosystems became degenerate, and formerly abundant resources became scarce or were no longer available. Survivors turned to other sources for sustenance, including some not on the original menu."

[This is a quote "In Six Days", edited by John F. Ashton, P.71]

In God's original plan, beneficial relationships between living organisms were, no doubt, intended. However, viruses, fungi, bacteria, and parasites are all microscopic and those that stand as aggressors would love to eat us if they could. Drugs can do so much but they are unable to do it all.

We can be thankful for the immune system which is explained in Michael J. Behe's book, *Darwin's Black Box*, chapter six – "A Dangerous World." Behe also demonstrates that each feature poses massive challenges to the Darwinian step-by-step evolution. Not only do these features have to be set in place and ready to go but the parts have to interact with each other. He reminds us "we must not automatically assume the different parts of the immune system are physical precursors of each other" (P.138).

In other words, every step in this highly complex defense system, before it is automated as an active force against microscopic invaders, has to be accounted for by some mechanism. This system must recognize the difference between invading bacterial cells and regular blood cells. It must be able to differentiate between connective tissue and the viruses which are marked for destruction.

A Brief Description of the Immune System:

* The immune system is a group of cells in the blood and a body fluid called lymph.

* Several types of white blood cells move to attack an invader which enters the body.

* One type of white blood cells moves to scoop up the invader. Along with the helper T-cells, they release chemicals that fight the invader.

* T-cells activate thousands of B-cells but they have to find the right one to release antibodies that will take care of the intruder. The B-cells, in their release of antibodies, will cause the invaders to clump together to be easily scooped up. Antibodies work against viruses by destroying the diseased cells. Helper T-cells activate killer B-cells.

* T- and B-cells called memory cells, remember each new germ. The memory cells trigger a defense against an invader that shows up a second time.

The bottom line is this – the immune system is irreducibly complex. That is, it must be in place and ready to function all at once. This system could not be built up over a long period of time. Behe says, "A cell hopefully trying to evolve such a system in gradual Darwinian steps would be in a quandary" (P.125).

Behe reaches this conclusion in his sixth chapter:

"Whichever way we turn, a gradualistic account of the immune system is blocked by multiple interwoven requirements. As scientists we yearn to understand how this magnificent mechanism came to be, but the complexity of the system dooms all Darwinian explanations to frustration."

P.139

David Quammen lists Biochemistry as one of the supporting data for Darwinian evolution. Apparently, Quammen has not read *Darwin's Black Box*. Quammen, in describing his visit to the office of the evolutionary biologist, Douglas Futuyama, writes: "I arrived carrying a well-thumbed copy of his own book ... *Science on Trial: the Case for Evolution*. Perhaps, if Quammen carried around a well-thumbed copy of Behe's *Darwin's Black Box*, then he might change his thinking about Darwinian evolution.

One more item to cover as David Quammen makes the outlandish statement in page 35 of the *National Geographic*, November 2004:

"Laboratory mice ... serve as research models because, sharing our mammalian ancestory, they also share a large proportion of our DNA." P.35

So far, in this work, we have investigated the various ways that organisms are supposed to share an evolutionary, mammalian ancestry:

* Tree of Life – The tree demonstrated just the opposite of mammalians sharing their animal ancestry. Twigs and branches are not connected to the base of the tree. There are too many dots which show disconnecting patterns in nature. Darwin wanted us to fill in the blanks with our imagination as we saw no evidence for a sequential order.

* Fossil Record – Rather than showing the ancestral relationships of mammals, the record is filled with gaps. To bridge these enormous gaps, there are simply no intermediates or transitional forms to be found. And yet, we are supposed to accept the outrageous doctrine of Darwinism.

* Embryological Research - will not prove to be suggestive of inheritance from a common ancestor (See APPENDIX II).

* Mutations – were not indicative of showing animal ancestory.

* Gene Recombination's – in mammals, proved that small changes could take place within "kinds" of animals rather than demonstrating each "kind" was linked to another "kind" by animal ancestory.

* Vestigial Structures – which we studied in depth, did not "prove" a sharing of mammalian ancestry.

* HOMINOID CHARTS - failed to show how man and monkey share a common ancestry. Although, this was not a complete study, enough was stated about charts to give a person the general idea as to their deficient nature. Evolutionists are turning away from charts with broken lines [the missing species from the fossil record] to CLADISTICAL CHARTS with their curved patterns of attractive colors. "Cladistics" is a new method which is derived from the Greek word meaning "branch." A cladist (someone who uses the cladistic approach) places all his confidence in the similarity of living things. His interpretation of species and their ancestral relationships to one another are still based on Darwinian classification. In other words, Darwinists are retreating to the old argument: homology serves as evidence for common ancestors. Old reasoning comes dressed in new garments but no matter how colorful these garments might be, they cannot conceal the decayed theory beneath. The main drawback in using the cladistical method, it emphasizes the discontinuous appearance of classes of living things. The order of nature is not sequential and cladistics does not "prove" ancestral relationships of species. The cladistical method is the same as any other method which attempts to classify species. When species are regarded as originating from a common ancestor, the taxonomist is talking about origins that exist only in his mind and not in the reality of nature.

Michael Denton writes:

"Small wonder then that the evolutionary community is viewing cladism with a growing sense of unease ..."

Evolution: A Theory in Crisis

P.139

Some evolutionists regard the practice of using cladistics, as "fools' errand."

David Quammen, in a final gesture and last-ditch effort, directs our attention to the mouse and how, through genetics, it shares a common ancestry with man.

Rather than, once again, going into the supposedly molecular "evidence" for the relationship between mammals, allow me to quote Michael

Denton who, I assume, is more qualified to make comments than David Quammen or myself. Michael Denton is an Australian molecular biologist and medical doctor who is best known for his biological research. He writes:

"There is little doubt that if this molecular evidence had been available one century ago it would have been seized upon with devastating effect by the opponents of evolution theory like Agassiz and Owen, and the idea of organic evolution might never have been accepted."

Evolution: A Theory in Crisis

Michael Denton

Pp.290-291

David Quammen, in his article "Was Darwin Wrong?-NO" put his trust in Darwin and the *Origin of Species*.

Michael Denton reminds us that Darwin knew nothing about the origin of new beings on this earth:

"The truth is that despite the prestige of evolutionary theory and the tremendous intellectual effort directed towards reducing living systems to the confines of Darwinian thought, nature refuses to be imprisoned. In the final analysis we still know very little about how new forms of life arise. The 'mystery of mysteries' – the origin of new being on earth – is still largely as enigmatic as when Darwin set sail on the Beagle."

Ibid. Pp.358-359 [Note: As far as Darwin's theory goes, "back to square one" for Darwinians]

Quammen, unquestionably, came to the wrong decision in following Darwin. His article in the *National Geographic*, has no doubt affected many thousands of people with the deleterious nature of his writing. This book has presented ample evidence in one section after the other why nature does not show a continual linkage of past and present life forms. The Darwinian concept of evolutionary ancestors has existed only in man's imagination and not in the facts of nature. Therefore, my positive response to David Quammen's question, "Was Darwin Wrong?" is a reasonable and suitable conclusion.

CHAPTER EIGHTEEN

MONARCH METAMORPHOSIS,
A DELICATELY SCULPTURED MIRACLE

I should think the very last thing an evolutionist would choose to do is direct our attention to the process of metamorphosis.

David Quammen writes:

" ... Embryology also concerns the immature forms of animals that metamorphose, such as the larval of insects" – Page 9 in the *National Geographic*, November, 2004.

There are two types of metamorphosis: the "incomplete" and the "complete." One insect, the Monarch, will be considered for its incredible transformation from caterpillar to butterfly – a "complete" metamorphosis.

Michael Denton is not a creationist but argues against the Darwinian concept of slow accumulation of beneficial mutations to account for metamorphosis of insects. Denton comments:

"An interesting example of a very widespread invertebrate phenomenon, the origin of which is in most cases difficult to account for in gradualistic evolutionary terms, is that of metamorphosis. Many invertebrates undergo a dramatic metamorphosis between the egg and adult form ... the transformation involves virtually the complete dissolution of all the organ systems of the larva and their reconstitution de novo from small masses of undifferentiated discs. In other words, one type of fully functional organism is broken down into what amounts to a nutrient broth from which an utterly different type of organism emerges."

Evolution: A Theory in Crisis

P.220

Denton claims that such an example as insect metamorphosis defies evolutionary explanation. In other words it is impossible to explain insect metamorphosis on the step-by-step process of Darwinian evolution.

David Quammen's contention is, animals that metamorphose show a homologous relationship and therefore demonstrate a common ancestory.

However, Denton explains:

"Insect metamorphosis provides many other examples of homologous organs and structures being arrived at by radically different embryogenic routes. The first stage of metamorphosis, shortly following the formation of the pupa involves what amounts to the virtual dissolution of all the organ systems of the larvae into a veritable soup of fragmented cells and tissues. This dissolution phase is quickly followed by an assembly phase"

Ibid. P.147

On the one hand, the organs and structures of insect metamorphosis are not homologous in the strict Darwinian sense (the organs and structures spring from different embryogenic routes) and, therefore, do not demonstrate a common ancestory. On the other hand, Denton describes metamorphosis for what it is – a true biological miracle. Metamorphosis is the process in animals by which a larva changes into an adult. This involves the radical change of internal and external body structures.

In simple steps: The fertilized egg becomes a caterpillar (quite unlike the adult butterfly). The caterpillar forms a cocoon and becomes a "veritable soup" (a jelly-like substance). From this "soup of fragmented cells and tissues" emerges this breath-taking monarch butterfly.

Imagine, if you will, what has just transpired! It is from this mass of jelly, THE WHOLE BODY IS REORGANIZED. Think of the extreme measures and incredible processes involved to take a green caterpillar with black and yellow bands, turn it into soup and transform this soup into a light brown with black borders and white dots, Monarch butterfly.

Does David Quammen really believe that an extraordinary event such as a caterpillar changing into a butterfly could possibly be the result of evolution? Does he believe that natural selection by slow events was responsible for molding the beautiful Monarch butterfly? The truth of the matter, there is no evolutionary evidence that can be produced to indicate how the remarkable plan of metamorphosis came about.

Evolution is not metamorphosis. The caterpillar did not evolve into a butterfly. The butterfly is a different creature from that of a caterpillar. There was nothing in the caterpillar that could evolve into wings. The caterpillar had mouthparts for chewing leaves but he had nothing that could assist him in shaping a long tongue for getting nectar out of flowers when he became a butterfly (Gish).

How did the caterpillar develop over long evolutionary periods and through one small step at a time into the butterfly form? How did the caterpillar, during this long interim, manage to reproduce itself; it doesn't lay eggs or reproduce at all. David Quammen should be attentive to the genetic mutations that are necessary to have the particular scenario of metamorphosis take place. The mutations or random changes in the molecular message of DNA would have to total more than the atoms in the universe before adding up to the evolutionary possibility of the processes involved in metamorphosis.

Dr. Duane T. Gish, Ph.D. in biochemistry has made this observation:

"But God, the Master Engineer of the universe, programmed a mass of jelly to develop into a delicately sculptured butterfly, with every-thing perfectly designed for a way of life entirely different from that of the caterpillar that dissolved into the jelly."

The Amazing Story of Creation, P.53

David Quammen is a strong supporter of natural selection. I have two observations which I would like my readers and Quammen to consider. These observations have to do with the Monarch butterfly and natural selection:

*   As winter approaches, the monarchs migrate by the thousands to the Southern United States and Mexico. Monarchs are capable of flying 2,000 miles from Canada to Mexico and back again to the Southern United States. The Monarch is without question, the champion of insect migration. The reason that they migrate is, they are not pre-adapted to winter survival. The 2,000 mile trip is, of course, not made by any one Monarch. The butterflies that set out from Mexico and California, die. This happens to several generations before they reach Canada. Even though nature is under the curse of death because of sin, the Master Designer's love is evidenced by His watch care over this world. The Monarchs, through migration, are led away from the cold and over-populated areas into new living quarters of warmer climate where forests are carpeted with delicate and freshly generated flowers. These flowers supply nectar for the Monarchs to drink. Along the way, each female lays up to two hundred eggs and the Monarch is able to generate its own species and assure itself a place in the future.

But what is truly amazing and remarkable about Monarch migration, the offspring in their return to California and Mexico go to the same forest that their great-grandparents left in February although they have never been there.

Andrew McIntosh, D.Sc. in mathematics, has written:

"This, of course, means that a remarkable system of information is bound up in the genetic coding of each butterfly, such that it 'knows' at what stage of the migrating cycle the group of butterflies is in. This information is passed on to each generation. Such a delicate mechanism shouts intelligent design! ... ... Furthermore, it has been established that magnetite has been found in the bodies of monarch butterflies ... indicating that they are able to orient themselves by sensing the earth's magnetic field. Their eyes are also sensitive to polarized light from the sun, again giving them a direction clue."

[This quote is "In Six Days" edited by John F. Ashton, Pages 167,168]

"... How could natural selection produce the Monarch butterfly, which transforms from a caterpillar to a butterfly with two compound eyes, each with 6,000 lenses, and a brain that can decipher 72,000 nerve impulses from the eyes?"

Design Features of the Monarch Butterfly Life Cycle

Impact 237 (March 1993): 2-3. Jules Poirier, Kenneth B Cumming

We should ask the question - Is this transformation marvel the result of natural selection or the design of the Master Engineer?

According to Darwin, the complexity of life, the design of the universe and the purpose behind every living system is the result of a blind random process called natural selection. The purpose of this book was to make clear that it is no longer tenable for empirical science to operate in the framework of orthodox Darwinism. Was Darwin Wrong? We must not only be able to answer this question in the *affirmative* but be aware that Quammen's *negative* answer to this same question is an affront to reason. Asking us to believe that we ourselves and everything that surrounds us is the result of blind chance, is a direct attack against common sense.

The apostle Paul warns men, the wrong way to study nature is from the perspective of "science falsely so called" (I Timothy 6:20). He also admonishes us to study nature the correct way ... "For the invisible things of Him from the creation of the world are clearly seen, being understood by the things that are made, even His eternal power and Godhead; so that they are without excuse" (Romans 1:20).

The apostle Paul, observing the wonderful works of design, concluded that man in his intellectual honesty must admit that there is a Designer or otherwise be without excuse. This text had its counter idea in the legal courts of Rome. It stated that if man enters the witness box without valid reason for his rebelliousness, he is without excuse. The same it will be when man stands before the judgment bar of God. Man's insolence in rejecting the revelations of the Great Designer, will hold him as being without excuse.

CHAPTER NINETEEN

DARWIN'S CLASSICAL CONCLUSIONS, NOT SO CLASSY

Darwin wrote many books but his two main works are:

The Origin of the Species

The Descent of Man

Also, he wrote his conclusions.

The first conclusion is from *The Origin of the Species* :

"There is grandeur in this view of life, with its several powers, having been *originally breathed by the Creator into a few forms or into one*; and that, whilst this planet has gone cycling on according to the fixed law of gravity, from so simple a beginning endless forms most beautiful and most wonderful have been, and are being evolved." [Emphasis, mine]

P.374

Comments: - The above quote is from the 6th edition of the 1859 book of Darwin, sold out on the first day of its publication. Said the wife of the bishop of Worcester after she learned of Darwin's publication, "Descended from the apes? My dear, let us hope that is not true, but if it is, then let us pray that it will not become generally known." Unfortunately, the theory of man's descent from the apes is still believed by a few, even in this present day.

Darwin writes of a Creator *breathing life* into a few forms or into one form. Where did Darwin get this thought if not from the Bible? The Bible is the only source that mentions the *breath of God* as being the awakening energy of life in the first human on this earth. The *one form* receiving his existence by breath was Adam, the first man. In the book of Genesis, chapter two and verse seven reads:

"The Lord God formed man from the dust of the ground and *breathed into his nostrils the breath of life*, and man became a living being."

Genesis 2:7

Talk about origins! – The first book of the Bible (Genesis) reveals the first form of human life on this earth (Adam); the first form of life's energy for Adam's life (Breath); the first source of energy for Adam's life (God).

The apostle Paul stood up in Athens and spoke about the *breath of life* in this manner:

"The God who made the world and everything in it is the Lord of heaven and earth and does not live in temples built by hands. And he is not served by human hands, as if he needed anything, because *he himself gives all men life and breath* and everything else … 'For in him we live and move and have our being.' As some of your own poets have said, 'We are his offspring.'"

Act 17:24-25, 28 NIV

How unfortunate for Darwin! He did not follow through with his thinking on God's creation of man in the book of Genesis. Man did not ascend from the apes, he was created. Man and ape did not ascend from an imaginary hominid ancestor. They were both created and with the God-given power, as were all men and animals, to perpetuate their own species.

Darwin compares evolution to the fixed law of gravity, proving that he could not discern the difference between a fixed law of science and the transient nature of his own theory. For Darwin, his theory was science.

Quammen wrote, "Darwin was a shy, conservative man" but anybody believing that an unproven theory is a fixed law of science, is not only self sufficient but demonstrates both an unrestrained and pretentious attitude.

The second conclusion is from *The Descent of Man* :

"The main conclusion arrived at in this work, namely, than man is descended from some lowly organized form, will, I regret to think, be highly distasteful to many."

P.919

Comments: - Not so much *highly distasteful* as it is *highly unscientific.* The myth of human evolution from so called "ape men" has been completely demolished. In *Bones of Contention*, a book assessing human fossils and written by Marvin L. Lubenow, who has researched the fossil issue for twenty-five years, has come up with overwhelming proof that humans existed contemporaneously or even before their alleged ancestors. How can you exist prior to your own grandfathers? This book is the most complete and accurate fossil critique of the "ape men." After reading the book, one is convinced that evolutionary scientists because of their faith, color the facts of their discoveries. They practice the art of self-deception and, therefore, lead the public into the same art. *Bones of Contention* takes the entire cladistical scheme of imaginative relationships of man and monkey and reduces it to the fairy tale that it has always been.

Continuing with Darwin's conclusions:

"For my own part I would as soon be descended from that heroic little monkey, who braved his dreaded enemy in order to save the life of his keeper, or from that old baboon, who descending from the mountains, carried away in triumph his young comrade from a crowd of astonished dogs – as from a savage who delights to torture his enemies, offers up bloody sacrifices, practices infanticide without remorse, treat his wives like slaves, knows no decency, and is haunted by the grossest superstitions."

Pp.919-20

Darwin, who did not believe in the Genesis account of creation, also passed over the Fall of Man which included its effects on the animals and plants. The apostle Paul informs us, in his letter to the church at Rome (8: 18-22), that animal suffering and plant degeneration were "subjected to frustration …"

Darwin, in his above statement, mentions the best in monkeys and the worst in man. He thinks that this is a fair and suitable comparison. But Darwin did not investigate the hideous habits of monkeys and apes as well as he did the wicked behavior of men.

In the *National Geographic*, March 1992, Boesh, in "A Curious Kinship: Apes and Humans" reports the following dark discoveries: [These thoughts might serve as a commentary on the animal degeneration which the apostle Paul speaks of in his epistle to the church at Rome]

*   "Apes once thought to live in a peaceable kingdom, are capable of killing one another" (P. 13).

*   "Infanticide has been observed among gorillas and chimps" (P.13).

*   "The Taï forest chimps also hunt cooperatively. A dominant male often leads a group of males and females to surround and kill red colobus monkeys" (Pp.23, 260).

[Boesch, an investigator of such matters, shrugs off any suggestions of cruelty] Perhaps, the Taï chimps are not aware of their cruelty. The red colobus is unable to define the word "cruelty" but certainly experiences pain and suffering while he is being eaten alive. Darwin, in a way, sat with those three imaginatory monkeys in a coconut tree and made the false claims that a monkey will not take some other monkey's life (maybe not with gun, club, or knife - but certainly with his own teeth).

* Some have said, "Warfare is a peculiarly human phenomenon and does not occur in other animals" but investigator Jane Goodall said, "In 1974 the chimp war started at Gombe, and by 1977 the smaller community was annihilated" (P.33).

So much for Darwin's observation of those savage humans as contrasted with the congenial acts of the heroic little monkey and the old baboon.

I was fifteen when I wrote in the margin of my copy of *The Descent of Man*. At the age of sixty-nine, I still think the same way. This is a take-off on Darwin's conclusion:

"For my own part I would as soon rejoice to be called a son of God and be descended from that heroic God-man who braved his dreaded enemy at the cross to save my life and become the keeper of my soul, or from that Old Testament Jehovah, who descending from heaven and uttering His law of love upon the mountain of Sinai, carried away in triumph his children from the crowd of evil agencies who are bent upon destroying moral laws and setting up the temples of pseudo-science."

Evolutionists claim humans are nothing but animals.

But Jonathan Wells writes:

"Despite the lack of evidence, the Darwinian view of human origins was soon enshrined in drawings that showed a knuckle-walking ape evolving through a series of intermediate forms into an upright human being. Such drawings have subsequently appeared in countless textbooks, museum exhibits, magazine articles, and even cartoons. They constitute the ultimate icon of evolution, because they symbolize the implications of Darwin's theory for the ultimate meaning of human existence."

Icons of Evolution: Science Or Myth?

P.211

And how is the materialistic view of humans, being no more than apes or monkeys, backed by science? Here is a short list of the supporting evidence used to prop up the mundane and humdrum speculation of man's ascent from the apes.

The fossil record –The fossil evidence found since 1912 bearing clear human-like or ape-like features have been so controversial that they have been called by the British anthropologist John Napier, "bones of contention." The story of ape to human evolution has not become clearer through fossil finds but has become more of a puzzle to the paleontologist's understanding. This is due to subjective reasoning that leaves a lot of room for interpretation and the anthropologists' preconceptions, which allows even more room for interpretation.

Ancestor-descendant relationships – Chapter eleven in this book ("The Orangutan's Visit to Dodger Stadium") has pointed to the scientific perils that the evolutionist encounters as he attempts to make comparisons between humans and apes. Weaknesses have been made known in the morphological or anatomical comparison, the biological comparisons, and the biochemical comparisons.

Henry Gee, Chief Science Writer for *Nature* wrote in his pessimism, "No fossil is buried with its birth certificate." Wells said, "It's hard enough, with written records, to trace a human lineage back a few hundred years. When we have only a fragmentary fossil record and we're dealing with millions of years – what Gee calls 'Deep Time' – the job is effectively impossible." Gee makes the observation, " To take a line of fossils and claim that they represent a lineage is not scientific hypothesis that can be tested, but an assertion that carries the same validity as a bedtime story – amusing, perhaps even instructive, but not scientific."

[Gee's quotes are found in *Icons Of Evolution Science or Myth*? Pp.220-221]

Paleoanthropology – is the science (?) that backs up the theory that humans made their ascent from the ape-line. But, as time goes by, Paleoanthropology is becoming *less and less* of a SCIENCE and becoming *more and more* of a MYTH. Wells cites a number of Anthropologists who have concluded the science of anthropology has been relegated to the "narrative framework of a folktale," "storytelling" by Paleoanthropologists, and as having "the form but not the substance of science." Wells points out those popular presentations of ape-to-man theories such as "Dawn of Humans" in *National Geographic* magazine, cover stories in *Time* or *Newsweek*, television specials on the Discovery Channel rarely mention disagreements among paleontologists, "but the public is rarely told the fossils have been placed into 'preexisting narrative structures' or that the story they are hearing rests on 'biases, preconceptions and assumptions.' It seems that never in the field of science have so many based so much on so little." Jonathan Wells give his assessment of such presentations –

"Whether the ultimate icon is presented in the form of a picture or a narrative, it is old-fashioned materialistic philosophy disguised as modern empirical science."

P.22

In coming to the conclusion of my remarks on Darwinism, I have sought to answer only the issues raised in the article of David Quammen and not to go any further than this. There have been issues not covered by Quammen which I would expect him to cover in additional articles of the near future. If he does, then I will respond – God willing – at the proper time. Yet, this presentation has covered the three important areas of: Biogeography, Paleontology, and Morphology. Embryology calls for a more detailed discussion and I have saved it for Appendixes 2 and 3. Under the subject of Paleontology, horse evolution is an extensive subject, and I have it listed under Appendix 1.

Careful readers of this book have seen *no air-tight proofs* in any of these areas which are used in an attempt to uphold the theory of evolution. The answer to Quammen's question, "Was Darwin Wrong?" has been responded to with a positive, "Yes."

Darwin's writings – *The Origin of Species* and *The Descent of Man* fail to sight one empirical example of macroevolution. My writing and research only bear out what evolutionists have known for quite some time – there is no mechanistic proof for evolution. There are minor modifications in species (microevolution) but no major modifications (macroevolution).

Roger Lewin, an evolutionist, makes the following admission that confirms the above sentiment and where evolutionary theory stands in the present scheme of history:

"An historic conference in Chicago challenges the four-decade long dominance of the Modern Synthesis ... one of the most important conferences on evolutionary biology for more than 30 years. A wide spectrum of researchers ... gathered at Chicago's Field Museum of Natural History under the simple conference title: 'Macroevolution'...
"The central question of the Chicago conference was whether the mechanisms underlying microevolution can be extrapolated to explain the phenomena of macroevolution. At the risk of doing violence to the positions of some of the people at the meeting the answer can be given as a clear, No. Species do in deed have a capacity to undergo minor modifications in their physical and other characteristics, but this is limited, and with a longer perspective it is reflected in an oscillation about a mean: to a paleontologist looking at the fossil record, this shows up as stasis."

Science, vol.210 (Nov.21, 1980)

"Evolutionary Theory Under Fire"

Pp.883-84

Many people left this meeting with the conviction that there is no evolutionary mechanism for macroevolution. They saw this meeting as the turning point in evolutionary theory. Others went back home convicted otherwise. I wonder what it was that persuaded them to not accept the clear answer, "No evolutionary mechanisms" to the alleged phenomenon of macroevolution! I would surmise, in their case, it was not a matter of conviction but a matter of *headstrong faith.*

APPENDIX 1

EVOLUTION OF THE HORSE –
FACT OR FICTION

THESE ANTIBIOTIC-RESISTANT STRAINS REPRESENT AN EVOLUTIONARY SERIES, NOT MUCH DIFFERENT IN PRINCIPLE FROM THE FOSSIL SERIES TRACING HORSE EVOLUTION FROM HYRACOTHERIUM TO EQUUS...

National Geographic, P.21

Comments: - David Quammen fails to demonstrate good judgment when it comes to mentioning the horse series. Rather than being cautious in alluding to a series that was, at one time, the archetype of progressive evolution, used in science textbooks, put on display in every museum, now has become the travesty of the scientific world, Quammen could not resist the temptation.

Stephen Gould refers to the horse series as the Exemplar of "Life's Little Joke." Gould writes in his book of over 1300 pages – *The Structure of Evolutionary Theory*, P.905:

" ... The line of horses, proceeding via three major trends of size, toes and teeth from dog-sized, many-toed, 'eohippus' with low-crowned molars to one-toed Equus with high-crowned molars ... still marches through our textbooks and museums as the standard-bearer for adaptive trending towards bigger and better .... Nonetheless, a speciation reformation in terms of changing diversity as well as anatomical trending tells a strikingly, and mostly opposite, story for the (horse) clade as a whole."

Unfortunately, "Life's Little Joke," has deceived countless people – including our youth – into believing this evolutionary hoax. In this appendix, it will be made crystal clear as to what Gould meant in writing, "in terms of changing diversity as well as anatomical trending tell a strikingly different, and mostly opposite, story for the (horse) clade as a whole." Gould, as a steadfast evolutionist, could not buy into straight-line evolution (progressive evolution or trending towards bigger and better). However, Gould was espoused to the doctrine of Punctuated Equilibrium.

Hank Hanegraaff defines this doctrine: "The long period of stasis is the portion of the process referred to as the period of equilibrium, and the interval characterized by rapid evolution is the punctuation – thus the term, punctuated equilibrium" (*The Farce of Evolution*, P.43).

But this theory – Gould's theory – has two problems:-

Firstly, Gould in explaining the lack of vertical transitional forms in the fossil record argues from the standpoint of silence – a classic, pointless argument.

Secondly, Gould's evolutionary concept defies the science of genetics by calling for a rapid and drastic alteration of gene structures in organisms otherwise known as having complex and stable genetic machinery.

Gould's teaching borders on Goldschmidt's absurd notion of the "hopeful monster theory." Richard Goldschmidt, a German geneticist, who believed there have been quantum leaps from one species to another such as "the first bird hatched from a reptilian egg." Whatever Goldschmidt had learned from the science of genetics, had been completely tossed out the window. Such a theory is contrary to the genetic equipment of a lizard which, as far as anybody knows, is 100% intent on producing another lizard (Gish).

Returning to horse evolution, I cannot bring myself to believe in Quammen's intellectual honesty in passing this bogus information (the horse series) on to the public. It's true; Quammen just barely mentions this series but gives the impression that it is *another evolutionary fact*. We will examine his brief remark by holding it under the light of true data discovered in the field of paleontology. Quammen's comment is calculated to mislead the unwary public into accepting what has become an imaginatory tale. In that the horse series has already been disclaimed by the scientific world, I can only believe that Quammen's motive is to hoodwink the public.

At on time, the horse series could be seen in the supposed order of their evolution. However, most museums have taken out the series because recent evidence has undermined its authenticity. Nonetheless, this illustration of progressive evolution continues to show up in text books and books on general science. This is the main reason for currently dealing with the subject and for presenting the complete story of the horse series as it once was. As I write about this subject, it will appear that *all* evolutionists believe in the horse series but you must understand that *most* scientists no longer accept the story of horse progression in the geological column. The other reason for dealing with the horse series is, of course, to inform readers who may not be familiar with any of the following account; readers who are familiar with the story but might value additional information; readers who know the details of this evolutionary scenario and have believed them to be true.

According to evolutionary paleontologists the horses and certain smaller types, generally considered as their ancestors, are found in different levels in the Tertiary deposits of the west, both on the plains and in Tertiary basins.

Thus, in a study of the evolution of these animals, it is said to be necessary to turn to the paleontologist for evidence. Evolutionary paleontologists maintain that such evidence consists of bones entombed in beds of sand and clay, most of them now hardened to rock and laid down in various parts of the earth during what is called the Tertiary period and part of the succeeding Quaternary period in which we are now living. They also claim that the order of succession of the horse fossils is determined by the order of the deposition of the bed in which the bones are found. It is logical to conclude that in a series of beds from the same locale, the one at the bottom was laid down first and the overlying beds were laid down in the order in which they appear one above the other. The bones found in these beds belonged to animals that lived and died about the time the beds were formed.

The horse fossils discovered by paleontologists, tell something about horse biology. Each bone occupied a certain known position in the skeleton of the animal with certain distinctive features that performed definite, well-known functions. Evolutionists claim that the work of tracing the development of the horse through the geological strata is relatively easy, especially the forms of the horse that lived in America. There are, they say, very extensive series of fossil remains of American horses, taken from beds that are piled in succession one upon another. Through these finds, the evolution of the horse is supposedly traced.

In the late 1970s, I resided at Rock Springs, Wyoming. From my car window, in certain regions, I remember the thrill of viewing wild horses galloping through the valleys and plains of this "cowboy" state. Stopping at roadside museums, I was not surprised to see the display of fossils with the usual evolutionary comments written beneath them. The recovered fossils were of the local flora and fauna. Each museum had its display of books and, at the time, the Bulletin by Michael W. Hager was popular. The author of *Bulletin* 54, the "Fossils of Wyoming" gives the following description of fossil horses that lived in North America:

"Horses lived in North America from the Eocene to the Pleistocene Epochs. Fossil horses from rocks of each succeeding epoch show a gradual increase in size, a reduction of toes from 5-1, and a change in tooth pattern.

"The oldest known horse, Hyracotherium (better known as 'eohippus,' the dawn horse) was about 20 inches high at the shoulders and had four toes on the front foot and three toes on the hind foot. Hyracotherium was a forest dwelling animal and had low-crowned teeth adapted to a diet of soft leaves ... ...

"Mesohippus, an Oligocene horse, was about as large as a collie dog. Its teeth were low crowned and, like Eohippus, fed on leaves and soft vegetation. The toes on the feet of Mesohippus were reduced to three in front and back and a vestige of the former toe is found as a small splint on the leg bone ... ...

"Merychippus, a Miocene horse ... was about as large as a Shetland pony. The middle toe of each foot was large and the side toes were short. The teeth of Merychippus were high-crowned and the enamel pattern was complex and set in cement, an adaptation for grazing on harsh prairie-grasses ... The legs adapted for running and the teeth adapted for grazing indicate that Merychippus and later horses took up life on the prairies and plains.

"Changes from Merychippus to the modern horse (Equus) involved mainly an increase in size, perfection of the grinding teeth, and the loss of toes ..."

December 1970, P.44

From the horse fossil finds, evolutionists have deduced the following data:

*   Through study of hundreds of specimens, a very complete picture of the sequence of gradual change in the horse lineage has emerged.

*   The history of the horse family is one of the clearest and most convincing demonstrations of the fact of evolution.

*   A number of fossils, from the oldest known horse to the modern horse can be laid in a series. From such a series, the fact of gradual change with time (evolution) becomes evident.

Thus far, the evolutionary scenario for the horse series seems logical in its presentation. Yet, is the synopsis truly accurate and irrefutable? Is it scientific in scope? Is the horse sequence a clear and convincing demonstration of the fact of evolution? The following information will attempt to answer such questions. Those who have leaned heavily upon the horse series may be surprised to find out the following facts and how these same facts have fictionalized evolutionary, gradual change of horses.

Again, to reiterate, the horses in the Tertiary deposits "constitute what the evolutionists regard as a perfect evolutionary series, with a gradual succession of changes in size, foot bones, and teeth. The series is admirably arranged as we view it in a museum, and looks quite convincing."

Fossils, Flood, and Fire

Harold W. Clark

P.166

But is evolution the only answer to the horse series? The creationist thinks not and for good reasons. Harold Clark makes a further observation:

"The series arrangement may, partially at least, represent mere taxonomic series rather than descent lines. That is, any group of similar animals … may naturally be expected to show variations in any one or more structural features. The fact that two or more such structural features show somewhat parallel series of changes does not prove them evolutionary. These features have an adaptive significance. An animal living in an environment that requires certain adaptive features in one organ or system will generally show corresponding adaptations in other systems of structural features." [Ibid. Pp.166-167]

Clark goes on to point out similar horse adaptations that the evolutionist Michael W. Hager has already emphasized in his *Bulletin* 54. Clark demonstrates how the three classes of horses were, by their size, teeth, and leg structure, adapted to the forest, plains, or marshes. However, Clark reasoned that it would be just as logical to conclude that the Horse Series represented ecological types rather than an evolutionary sequence.

Clark in mentioning that "forest horses" which were adapted for conditions between the two extremes of "marsh horses" and "plain horses" arrived at the following conclusions: "The three classes of horses, when viewed in the light of ecological principles, can be explained without the necessity of resorting to evolution ; the horses could have existed contemporaneously and could very well have occupied different ecological zones; the strata in which the horses were found do not indicate geological ages but stages of flood action (such as the biblical world-wide flood) wherein the strata was laid down in rapid sequence resulting in the burial of the various life zones; the whole idea of serial progression in horses is based purely on hypothetical concepts, the same as any other evolutionary theory. The horse series may be explained as taxonomic or ecological groups, or by considering the possibility of hybridization between closely related groups without having to accept the commonly offered evolutionary interpretation."

Thus far, the evolutionary theory of the horse series has been presented as well as the opposing creationist theory. We have looked at the Tertiary period and part of the Quaternary; have seen that the exact same geological and paleotological phenomena can be observed and yet the observers come to different conclusions. What enlightening evidences can be revealed which would render the thinking of one theorist to be more logical and scientific than the other theorist? It is now imperative to present solid FACTS which will militate against the evolutionary horse series and render it to be completely unscientific.

FACT ONE – The horse series cannot be demonstrated in the rocks (strata) by the law of superimposition. This law states that the older rocks, by definition, will always contain simpler animals than does recent rocks. Frank Lewis Marsh makes the following observations:

"These authorities in historical geology aver that the rocks contain abundant proof of evolution. They are apparently absolutely sincere in this conclusion. But the faith in an unproved theory which if is manifested in these workers as they pull together scattered fossils from all parts of the world and build them into arbitrary pedigrees of elephants and camels, just as in the case of the horses, is most astonishing to the uninitiated."

Evolution, Creation, and Science

P.275

In other words, the horse series does not appear as an evolutionary story in the vertical feet of stratified rock. The assumed fossil ancestors of modern horses are pulled together from many different areas and arranged in a series dictated by evolutionary theory rather than paleotological fact. Marsh gives his exceptional scientific perception: "Only the other day I was studying the fine collection of fossil horses on exhibition in Morrell Hall on the campus of the University of Nebraska. The following is a list of the members of this particular pedigree form 'youngest' to 'oldest,' with the region in which the fossil was found: Eohippus from Wyoming; Mesohippus from Sioux County, Nebraska; Miohippus from Oregon; Merychippus from Daws County, Nebraska; Pliohippus from Daws County, Nebraska; Plesippus from Idaho; and Equus excelsus from Sheridan County, Nebraska.If these fossil horses had been found in this order in successively younger strata, according to the law of superposition, then there would be some reason for considering that each larger and more specialized form had evolved from a smaller form before it. But since they had been pulled together from Oregon, Idaho, Wyoming and various counties in Nebraska, it is every bit as sensible to conclude that they were all living on the earth at the same time." [Ibid. Pp.272-273]

The only reason for arranging the horses in order from the Dawn Horse to the size of the modern horse is the assumption that evolution has occurred. Yet, if the strata where these horse fossils were discovered by paleontologists were super positioned, the geologists would have to increase the diameter of the earth 190 miles. The readers will note that the justification for this type of reasoning is heavy on theory and light on scientific analysis. Returning to FACT ONE, the horse series cannot be demonstrated in the rocks (strata) by the law of superposition. Otherwise, the total thickness of the stratified rocks would exceed 500,000 feet (=95 miles), which is impossible since the earth is only about 8,000 feet in diameter.

FACT TWO – Horses of the so-called horse series, at times, are out of order. Frank L. Marsh writes:

"When miners in Colorado found a hoof of a modern horse deeper in conformable strata than the bones of Eohippus, the 'freak of nature' of the schoolman appeared in modern transcendental form as a case of strata in the 'wrong' order, and nothing more came of the discovery. The only real reason   Eohippus is set ahead of Excelsus is not that their position in the rocks shows the former to be the older, but merely that the theory of evolution demands such a sequence. This is true for all other fossil pedigrees which are foisted upon the gullible public in our educational museums." [Ibid. P.287]

Any fossil appearing in a sequence that does not harmonize with evolutionary theories is immediately repudiated and labeled a "freak" of nature. It is claimed to be in the "wrong" order. Obviously, evolutionists refuse to weigh the significance of such evidence and will do anything to uphold their pet theory even if it means an artificial arrangement of paleotological material. The horse series is nothing more than an arbitrary arrangement of synthetic pedigrees. How unscientific it is, to palm off this false horse series on the public as genuine!

Astounding and revealing facts are discovered when North American horses are compared with South American horses. Fossil discoveries in South America make it crystal clear that these horses did not give rise to the one-toe from the three-toe as was generally believed by most evolutionists. In fact, the horse fossils discovered in South America are reversed when compared to the discoveries in North America. Duane T. Gish, an ardent creationist who has debated evolution at most major universities in the United States and Canada, makes this ascertainment concerning horse toes and feet:

"Do they not thus provide another nice, logical evolutionary series? No, not at all, for they do not occur in this sequence at all! ... in South America a one-toed ungulate gave rise to a three-toed ungulate with reduced lateral toes, which then gave rise to an ungulate with three full-sized toes. This is precisely the opposite of the supposed sequence of events that occurred with North America horses ...In the Rattlesnake Formation of the John Day County of northeastern Oregon; the three-toed Neohipparion is found with the one-toed horse, Pliohippus. No transitional forms between the two are found. In other cases 'primitive' species of a genus, such as those of Merychippus, are found in geological formations supposedly younger than those containing 'advanced' species."

EVOLUTION: the Fossils STILL say NO!

P.192-193

The above observations by Marsh and Gish are devastating to the theory of progressive evolution in the horse series. According to the fossil finds, horse evolution is non-existent. There are simply too many contradictions from the fossils, as they appear in the geological record, to advance the notion of progressive evolution in the horse series.

One further item needs to be addressed about the fossil discoveries in South America. Since the horse fossils of North America are in a sequence that is just the opposite of the sequence found in South America, evolutionists cannot have it both ways to keep their theory intact. Reverse transitions of horse fossils are devastating to the theory of progressive horse evolution. The horse genera simply did not evolve in two different directions at the same time. Let us note a few more problems in the so-called evolution of North American horses:

In Oregon, the three-toed Neohipparion is found with the one-toed horse, Pliohippus and there are no transitional forms between them. A fair question is this: How can Pliohippus (a one-toed horse) of 9 million years ago be found co-existing with a three-toed Neohipparion that supposedly pre-dated Pliohippus by many millions of years and was supposed to be its progenitor and already extinct? According to evolutionary theory, both horses could not be found in the same geological level. The creationist would have no problem as he would expect both horses to be living contemporaneously, with no long ages between them and with no evidence of progressive or linear evolution.

Another problem is worthwhile mentioning: "primitive" species of a genus such as those of Merychippus are found in strata that are supposed to be younger than strata containing "advanced" species. Merychippus, according to the evolutionary time scale, goes back to the Miocene epoch about 22 million years ago. This problem is major in scope. How can Merychippus be found in a geological formation that is younger than the strata in which "advanced" species are found? How can the "fathers" be younger than their evolutionary "offspring's age"?

The above mentioned problems alone, falsifies the concept of linear descent in horse evolution. No small wonder the "horse series" is being disregarded in scientific circles! The idea has no place in the realm of true science. It should be clear that the idea widely promoted by evolutionists that the three-toed horses evolved into one-toed horses is definitely not supported by the paleotological evidence.

FACT THREE – Horses of the so-called "horse series" often appear contemporaneously in the fossil record. Duane T Gish Comments:

"Three-toed horses and one-toed horses commonly coexisted together in North America. An example of this is found in Northeastern Nebraska. There is a remarkable fossil graveyard with a great variety of birds, reptiles, and mammals. The deposit was discovered by Michael Voorhies, curators of vertebrate paleontology at the Nebraska State Museum ... Among the rich mammalian fauna were found species of horses."

EVOLUTION: the Fossils STILL say NO!, P.193.

This graveyard is most remarkable to the evolutionist for in it he finds species of horses that have three-toes and one-toe. The reader may be quick to discern that when such horses coexist, the theory of evolutionary progression of the horse must be abandoned. There can be no evolutionary advancement from three-toed to one-toed horses since the remains are found at the same geological level. Even though this fact has been well substantiated, still the idea of evolutionary advancement is still being published in some text books. How can any honest investigator possibly believe that one horse evolved from the other when they were actually living at the same time level?

I have seen many evolutionary books with drawings and illustrations of various horses arranged in a series shown in their various geological time zones. The evolutionary time period for the horses goes back 55 millions of years ago. The earliest time zone goes back to the Eocene and works its way up the geological column through the Oligocene, Miocene, Pliocene, and finally the modern horse is found in the Pleistocene or most recent strata. We already have knowledge that the horse series cannot be demonstrated in the rocks (strata) by the law of superimposition; horses are often out of order. Because horses of the series are living at the same time level, we must conclude that the arrangement of various horses shown in geological time zones is contrary to paleotological evidence.

For some period of time, I have been familiar with the fact that horses from the "horse series" have been found at the same time levels but I was surprised to find out *all* the "horse series" have been discovered occupying the *same time level.* Walter R. Barnhart, in his thesis for Master of Science, 1987, Pp.148-150, came to this conclusion in his study of the horse genera: - "He reported that the geological range of taxa indicates that all the genera of horses are found in the Miocene and thus the geological progress required by evolution does not exist"(Reference in Gish, *EVOLUTION: the Fossils STILL say NO!*, P.195).

Any paleontologist knowing the above mentioned facts should abandon the concept of straight-line evolution for the horse series. There are no true orthogenetic series of horses in the geological column. All the genera of horses are found in the Miocene. Some evolutionary geologists claim that the horse series extends over a period of time covering 58 million years. Yet, as we have just learned, all the horses lived together in the Miocene period – a time period believed to be but 12 million years in length. This does not mean that horses did not live in other so-called evolutionary time periods but it does mean that in just the Miocene alone, complete genera of these horses have been recovered.

FACT FOUR – Hyracotherium (Eohippus) was not really a horse.

Richard Owen, the great British paleontologist, described the genus Hyracotherium in 1841. Scientists realized that Owen's older discovery represented the same animal that the paleontologist, Marsh, later named Eohippus. Under the rules of taxonomy, Owen's description takes precedence over Marsh's Eohippus which means "dawn horse." Even though Eohippus or Hyracotherium was not like modern horses, this creature was selected by Marsh and other scientists to stand at the base of the so-called progressive evolution of the horse series. This was done in spite of the fact that the "dawn horse," was morphologically, and in its environment, unlike modern horses.

Amazing, that scientists can select a creature to start off the horse series with no scientific basis for such a move! And then there was Thomas H. Huxley, who accepted the publication of Marsh's studies and, after a lecture in New York, solidly entrenched the Eohippus scheme both into popular circles and in scientific status (Gould). Huxley was a man of great intelligence but intelligence alone should not be the criterion for taking liberty in an attempt to establish mere theory into scientific fact.

G. A. Kerkut expresses his doubt whether Hyracotherium was related to the horse:

"In the first place it is not clear that Hyracotherium was the ancestral horse. Thus Simpson (1945) states: 'Matthew has shown and insisted that Hyracotherium (including Eohippus) is so primitive that it is not much more definitely equid than taprid, rhinocerotid, etc. but it is customary to place it at the root of the equid group.'"

Implications of Evolution, P.149

Where was the objectivity of those scientists who constructed the phylogenitic tree of the horse in the first place? The "horse" on which the entire family tree of the horse rests was not a horse at all. In this case, "a horse is *not* a horse, of course, of course."

Apparently, some paleontologists feel that they are at liberty to do whatever makes their theory look good. If placing Eohippus at the beginning of the horse series to make the scheme look more progressive because of the four-toes, then why not do so. No matter that the four-toed Eohippus does not look the least bit like a horse; no matter that Eohippus fossils have been found in the same strata alongside two types of modern horses.

At a glance, it is easy to see that science is not always objective and especially evolutionary science. Eohippus was a small animal and placing it at the beginning of the horse series, gives the allusion of simple to complex evolutionary progression. Yet, it really does not prove a thing for evolution by its presence. Luther D. Sutherland has this to say:

"The series shown in museum displays generally depicts an increase in size, and yet the range in size of living horses today from the tiny American miniature ponies to the enormous Shires of England is as great as that found in the fossil record. It is no wonder that Dr. Eldredge called the textbook characterization of the horse series 'lamentable.'"

Darwin's Enigma, P.82

On February 18, in 1870, Thomas Henry Huxley "gave his annual address as president of the Geological Society of London and staked his celebrated claim that Darwin's ideal evidence for evolution had finally been uncovered in the fossil record of horses – a sequence of continuous transformation, properly arrayed in temporal order."

Bully for Brontosaurus

Stephen Jay Gould, P.169

However, Huxley was wrong. There is no "sequence of continuous transformation." The fossil record of horses is not "Darwin's ideal evidence for evolution." Sunderland makes this statement concerning the so-called progressive series of horse fossils:

"Nowhere in the world are the fossils of the horse series found in successive strata. When they are found on the continent, like the John Day formation of Oregon, the three-toed and the one-toed are found in the same geological horizon (stratum). In South America, the one-toed is even found below the three-toed creature. And when other structures besides toes are considered, the picture does not look so impressive. For example, the four-toed Hyracotherium has 18 pairs of ribs, the next creature has 19, then there is a jump to 15, and finally back to 18 for Equus, the modern horse. The sequence requires arranging Old World and New World fossils side by side, and there is considerable dispute about the order in which they should be arranged. One specialist says, 'The story depends to a large extent upon who is telling it and when the story is being told.'" [Op. cit., p.81]

Added to the points we have already encountered, it appears that there is a further monkey-wrench to be thrown into the progressive evolution of the horse. The number of ribs that goes up and down in the geological column indicates that there were different kinds of horses in the past and has nothing to do with their evolutionary or phylogenitic relationship. As the number of ribs did not gradually increase, it is easy to discern that this is not progressive evolution and, in fact, the non-transitional forms testify against the truth of evolution and only support speciation.

Duane T. Gish gives his comment:

"Horses used to be evolutionists' favorite example of their theory. However, this is no longer true. Although I have lectured and debated on most major university campuses in the United States and Canada and in more than 20 foreign countries, I have rarely heard evolutionists mention the fossil record of horses as evidence for evolution. It is now recognized by most paleontologists, that the fossil record of horses is inconclusive when it comes to evolution, and, therefore, evolutionists no longer want to talk openly about it."

The Amazing Story of Creation from Science and the Bible

P.64

In 1945, the paleontologist George Gaylord Simpson (a neo-Darwinian) wrote that the evolution of the horse should be viewed as a branching tree pattern, not a straight-line evolution (the so-called orthogenetic sequence). The creationist also uses the bush metaphor to diagram the horse kind. Of course, there is always the danger of oversimplification on the part of the evolutionist as well as the creationist. The horse fossils form a complex pattern throughout the geological column but, nevertheless, the points listed in the summary remain salient since this section has made it clear that even in the so-called horse series, no one horse descended from any other.

SUMMARY

The argument for horse evolution appears to be not only weak but non-existent. The following points have already been evaluated but they stand as brief reminders of the information that we have gleamed from this appendix:

*   From the standpoint of logic, it is just as feasible to view the horse series as representing ecological types as it is to see them as an evolutionary sequence.

*   The horse series can not be demonstrated in the strata by the law of superposition. The assumed fossil ancestors of modern horses have been pulled together from many different areas.

*   Horses of the so-called horse series have been proved to be out of order in the geological column. Since this is so, progressive evolution is an invention and nothing more.

*   Eohippus, the first horse in the so-called evolutionary sequence is not a horse at all. To place it first in the series is to do so without any scientific reason.

*   Horses of the alleged evolutionary sequence appear contemporaneously in the fossil record. In fact, all of them are found in the Miocene period. This fact militates against any possibility of horse evolution. What was once a strong point for the doctrine of Darwinism is rendered totally powerless.

*   The number of ribs that goes up and down the horses as they appear in the geological column is a crucial testimony against progressive evolution.

A SHORT HISTORY OF EMBRYOLOGY -
THE TRUE AND FALSE STAGES

On page 13 of the *National Geographic*, November 2004, Quammen lists embryology as one of the main evidences of evolution.

EMBRYOLOGY TOO INVOLVED PATTERNS THAT COULDN'T BE EXPLAINED BY COINCIDENCE. WHY DOES THE EMBRYO OF A MAMMAL PASS THROUGH STAGES RESEMBLING STAGES OF THE EMBRYO OF A REPTILE? WHY IS ONE OF THE LARVAL FORMS OF A BARNACLE, BEFORE METAMORPHOSIS, SO SIMILAR TO THE LARVAL FORM OF A SHRIMP? WHY DO THE LARVAE OF MOTHS, FLIES, AND BEETLES RESEMBLE ONE ANOTHER MORE THAN ANY OF THEM RESEMBLE THEIR RESPECTIVE ADULTS? BECAUSE, DARWIN WROTE, "THE EMBRYO IS THE ANIMAL IN ITS LESS MODIFIED STATE" AND THAT STATE "REVEALS THE STRUCTURE OF ITS PROGENITOR."

We should note some other statements recorded by Charles Darwin in *The Origin of Species*. They are significant to the subject of embryology and especially how scientists viewed the embryo of man and its resemblance to the lower animals. "Man is developed from an ovule, about the 125th of an inch in diameter, which differs in no respect from the ovules of other animals. The embryo itself at a very early period can hardly be distinguished from that of other members of the vertebrate kingdom ...

"After the foregoing statements made by such high authorities, it would be superfluous on my part to give a number of borrowed details, showing that the embryo of man closely resembling that of other mammals ...

"Even at a later embryonic period, some striking resemblances between man and the lower animals may be observed ...

"With respect to development, we can clearly understand, on the principle of variation supervening at rather late embryonic period and being inherited at a corresponding period, how it is that embryos of wonderfully different forms should still retain, more or less perfectly, the structure of their common progenitor. No other explanation has ever been given of the marvelous fact that the embryos of a man, dog, seal, bat, reptile, etc., can at first hardly be distinguished from each other."

Pp.398, 400, 411

Comments: - Before Darwin wrote his *Origin* in 1859, there were many biologists at the beginning of the nineteenth century who recognized in their study of the development of embryos "the similarity in form and external detail even in the cases of widely divergent 'species' of animals. Equally obvious in embryos was the general conformity to a basic plan manifested by the various organs and organ systems."

Evolution, Creation, and Science

Frank Lewis Marsh

P.244

In discussing embryonic development, Darwin simply reiterated the findings of these "high authorities" in *The Origin of Species*, pages 398-411. Darwin, like his contemporaries, was of the opinion that the embryo of man closely resembles that of other mammals.

As early as 1828, Karl Ernst von Baer, called attention to the fact that the *general* of a large animal group expresses itself earlier in the embryo than the *special*. What does this mean? On page 122 of *The Revised & Expanded Answers Book*, there is found an elucidating explanation:

"Namely, the *general* features of a large group of animals appear earlier in the embryo than the *specialized* features. Less general characters are developed from the more general, and so forth, until finally the most specialized appear. Each embryo of a given species, instead of passing through the stages of other animals, departs more and more from them as it develops." [Emphasis, mine]

Biologists soon christened this principle of differentiation: "Von Baer's laws." Von Baer was the leading embryologist of the nineteenth century who discovered the mammalian egg cell in 1827. In 1828, he published the greatest monograph [According to Stephen Gould] in the history of the field – *The Developmental History of Animals*.

Von Baer argued that "a human embryo grows gill slits not because we evolved from an adult fish but because all vertebrates begin their embryological lives with gills. Fish, as 'primitive' vertebrates, depart least from this basic condition in their later development whereas mammals, as most 'advanced' lose their gills and grow lungs during their maximal embryological departure from the initial and most generalized vertebrate form."

I Have Landed

Stephen Jay Gould

Pp.317-18

Before continuing on with the brief history of those scientists who studied embryological development, there are two things that we should note about Von Baer: his theories of natural history allowed for limited evolution among closely related forms but not a major transformation between major groups and consequently held no sympathy for Darwin's mechanistic views of evolutionary causality. He did not accept the doctrine of recapitulation but unfortunately his concept of gill slits promoted the very theory he was against.

It was Fritz Müller (1821-1897) who ran with Von Baer's theory and gave it the recapitulation twist. Müller claimed that the evolution of the species is a "historical document" displayed in the development of each new individual. His paper "Für Darwin" was published in 1864 and inspired Haeckel (1834-1919) to make this recapitulation principle for the origin of life into his biogenetic law. This biogenetic law became the crux of the theory of recapitulation.

Recapitulation briefly defined is "Ontogeny repeats phylogeny." Recapitulation amplified is "The theory that holds that in the course of growth and differentiation from the fertilized egg, to the adult stage, higher forms of life pass through a series of stages closely similar to the adult stages of their ancestors." In other words, during its life history every animal climbs up its family tree. However, Haeckel, under his definition of "biogeny" (his own coined word) wanted this law to be more inclusive by having it stand as a law for the entire history of evolution.

We now come to the heart of the matter as to why such a theory as recapitulation caused uproar in the camp of both the creationist and evolutionist. A hundred years ago, the recapitulation theory was accepted by most (but not all) evolutionists. This theory went by other names: The law of morphogenesis; doctrine of parallelism; morphogenetic theory; and repetition theory. Controversies arose from the uncertainties of the evidence. It was not positively confirmed that embryonic development confirmed genetic relationships. Haeckel was quick to indiscriminately apply this controversial theory of recapitulation into a law and not only did he make recapitulation into a law but he attempted to establish this dictum by the most nefarious and corrupt ways.

In elaborating upon the brief history of Ernst Haeckel, you will immediately understand why his dishonesty brought consternation and concern to the creationist camp. If T.H. Huxley has been called the "bulldog" of Darwinian evolution then Ernst Haeckel has to be named the "general." Although Haeckel was considered an "apostle" of Darwinian evolution, it must not be assumed that he preached an exact Darwinian interpretation. Rather, he was a combination of three men — Goethe, Lamarck, and Darwin (Gould). In 1874, he wrote in his most popular book, *The Evolution of Man:* "… Evolution and progress stand under the bright banner of science … under the black flag of hierarchy, stand spiritual slavery and *falsehood* …" [Emphasis mine]

You would think that Haeckel should be the last one to speak of falsehood when his work on the *Evolution of Man* was founded upon lies and dishonesty. *The Revised & Expanded Answer Book* makes this comment:

"L. Rütimeyer, professor of zoology and comparative anatomy at the University of Basil corroborated with William His Sr., a famous comparative embryologist. These scientists showed that Haeckel fraudulently modified his drawings of embryos to make them look more alike."

P.118

In *The Farce of Evolution*, P.95, Hank Hanegraaff has this to say about Haeckel:

"His dishonesty was so blatant that he was charged with fraud by five professors and convicted by a University court at Jena. His forgeries were subsequently made public with the 1911 publication of *Haeckel's Fraud and Forgeries.*"

Ernst Haeckel was a German naturalist who had a great deal of energy. This strength would be exhibited in his volumes of technical, taxonomic descriptions. However, he took much liberty in imposing his theoretical beliefs upon the public. The shame about all this: his books along with their inaccuracies appeared in all major languages and he was more influential than Darwin or Huxley in convincing people throughout the world about the validity of evolution.

Let us now consider the artistic nature of Haeckel whose talents eventually led him down the pathway of deceit. Haeckel was a skilled artist with the brush and would often prepare his own illustrations for his technical monographs and scientific books. From the beginning, critics recognized the master naturalist and competent artist took systematic license in improving his specimens to make them more symmetrical and artistic. However, in doing this, he often led his reading audiences away from authentic nature to imagined nature in order to accommodate his theories. He altered the system of created things through his misleading idealizations.

A long time ago, in the year 1952, I received my high school course book for biology. On the inside front and back covers there appeared the famous chart of the embryological development of eight different vertebrates. At the time, I was unaware that Haeckel drew these vertebrates for a 1903 edition of his popular book, *The Evolution of Man*. It was only much later in life that I was finally made aware that Haeckel exaggerated the similarities of the earliest stages with gill slits in order to push his evolutionary law of recapitulation.

The eight vertebrates drawn by Haeckel were the Fish, Salamander, Tortoise, Chick, Hog, Calf, Rabbit, and Man. Stephen Jay Gould in, *I Have Landed*, recounts Haeckel's phony drawings. Gould first describes the latest depicted stages of the distinctive features of adulthood in Haeckel's embryonic drawings such as the tortoise's shell and the chick's beak. But, as Gould points out, Haeckel draws the earliest stages of the first row, showing tails and gill slits just under the primordial head, as virtually identical for all embryos whatever their adult destination. Gould also charges that Haeckel could thus claim that near identity marked the common ancestry of all vertebrates. Under the theory of recapitulation, embryos pass through a series of stages representing successive adult forms of their evolutionary history.

Gould writes:

"He also, in some cases – in a procedure that can only be called fraudulent – simply copied the same figure over and over again … … But these early embryos also differ far more substantially , one from the other, than Haeckel's figures show. Moreover, Haeckel's drawing never fooled expert embryologists, who recognized his fudging right from the start."

Pp. 309-310

I have a question that needs to be answered. I am certain that Gould in writing "Haeckel's drawing never fooled expert embryologists who recognized his fudging [too nice a word] right from the start." But I'm not an embryologist – let alone an expert in this field. How is it that these phony drawings from a book in 1903, appeared in my biology book of 1952? We can be certain, for the simple reason of pushing one of the so-called proofs of evolution – recapitulation. Stephen Gould assures us, Haeckel's famous argument of "ontogeny recapitulates phylogeny" has been disproved by science long ago. Nevertheless, the argument is still being taught and that is why I am issuing a warning to my readers.

Continuing, Gould gives reasons why we are still seeing Haeckel's charts in the standard student textbooks of biology: - Gould claims that Haeckel's inaccurate drawings entered the student textbooks because the books are considered "quasi-scientific" literature. In other words, high school and college text-books are not considered expert scientific literature. But what educational boards approved these text-books? And certainly, the teachers of biology who allow such material to come into their classrooms, must have a scholarly opinion of the text-books and their "noted inaccuracies" and falsifications.

Gould adds "But shortcuts tempt us all, particularly in the midst of elaborate projects under tight deadlines." Should it be understood that pressure automatically frees educators from the responsibility of making sure everything that makes for a bad education, is not passed on to the students? And what about the old but certain concept, students of today are the leaders of tomorrow! Is anybody concerned what the future leaders are being taught? No small wonder that fraudulent concepts are being perpetuated! Because Haeckel was a spokesman for Darwin and he also held high, professional credentials, text-book authors borrow his famous drawings of embryonic development probably, according to Gould, "quite unaware of their noted inaccuracies."

Gould assumes too much from the text-book authors. First of all, he assumes the authors considered every Darwinian spokesman must be genuine and that is why they were unaware of Haeckel's inaccuracies. Secondly, we can't help but wonder if Gould actually believed his assumption that text-book authors are completely "unaware" of the history of Haeckel and his phony drawings. I propose that anyone interested enough to write a textbook is interested enough to know the history of Professor Haeckel and his background. However, we must be aware of the exceptional cases. For example, Jonathan Wells cites a well known textbook-writer Douglas Futuyma. Futuyma is a professional biologist and author of a graduate-level 1998 textbook, *Evolutionary Biology*. Wells writes:

"Futuyma ... did not know about Haeckel's faked drawings – a confession of ignorance not likely to inspire much confidence in the quality of our biology textbooks. But now he knows that 'Haeckel was inaccurate and misleading,' and he said he would take this into account in future editions of his book."

"Icons Of Evolution Science Or Myth? [Sub-title: Why much of what we teach about evolution is wrong]

P.107

Gould and creationists as well, are concerned about the textbook problem but from two different standpoints. Gould gives his attention to this matter because he is concerned that incorrect information will reflect poorly upon the evolutionary doctrine of Darwinism which he holds to be valid and true. The creationist, on the other hand, is concerned because another so-called evolutionary "proof" will become fixed in the minds of young people who are being taught a doctrine which creationists accurately deem as invalid and false.

This brings up Gould's final point for text-books and their inaccuracies:

"... but then rationalized with the ever-tempting and ever-dangerous argument, 'Oh well, it's close enough to reality for student consumption, and it does illustrate a general truth with permissible idealization.'" Gould responds to this dangerous argument in the following way:

"I am a generous realist on most matters of human foibles. But I confess to raging fundamentalism on this issue [Gould, must be angry!] The smallest compromise in dumbing down by inaccuracy destroys integrity and places an author upon a slippery slope of no return."

I Have Landed, P. 311.

Talk about destroying integrity! – Jonathan Wells writes:

"So Gould blames the textbook writer, while the textbook writer pleads ignorance...

"But it was Futuyma who mindlessly recycled Haeckel's embryos in several editions of his textbook, until a 'creationist' criticized him for it. And it was Gould who (despite having known the truth for over twenty years) kept his mouth shut until a 'creationist' (actually, a fellow biologist) exposed the problem. And all that time, Gould was letting his colleagues become accessories to what he himself calls 'the academic equivalent of murder'"

Icons of Evolution

We can agree with Gould but only up to a certain point- such an author would be placed on slippery grounds. However, can we really believe, in the first place, Haeckel's drawings "illustrate a general truth?" and, in the second place, the drawings and illustrations were done "with permissible idealization"? For the sake of making a point, Gould placed the above words into the mouth of his teacher example but … … those words were, in fact, expressing Gould's own personal innervations – the "ever-tempting" rationalization, voices his own feelings.

Dr. E. Blechschmidt reveals some of his frustration with the persistence of this myth ("biogenetic law"). He doesn't appear to see "general truth" in any of it:

"The so-called basic law of biogenetic is wrong. No buts or ifs can mitigate this fact. It is not even slightly correct or correct in a different form. It is totally wrong."

The Human Body: An Intelligent Design

Alan L. Gillen, Frank J. Sherwin lll, Alan C. Knowles

P.33

When Gould wrote about recapitulation, even though he disagrees with the theory, it is obvious that he holds to part of it. Hank Hanegraaff in his marvelous book, *The Farce of Evolution* and on page 94 writes: "Gould proceeded to write an entire book to demonstrate that it is still 'one of the great themes of evolutionary biology.'" The book is entitled, *Ontogeny and Phylogeny*. It was Gould's first technical book and following my reading of it, I was able to form two possible conclusions: - Haeckel's dictum "ontogeny repeats phylogeny" is shattered; it is also made clear that Haeckel's forgeries and doctrine become irrelevant to the so-called validity of evolution's mechanisms. In other words, Gould still directed his readers towards the science of embryology and how, at the root of this science, it seeks to explain, through the great pageant of evolution, the marvelous diversity of life's forms.

Hank Hanegraaff writes:

"While admitting that the problems with Haeckel's recapitulation theory are myriad, Gould says he does not want to throw the baby out with the bath water." [Ibid. page 200]

Actually, Gould was not the first scientist to use this metaphor. M.F. Guyer in his *Animal Biology* of 1941, P.533, wrote:

"Some zoologists repudiate the whole doctrine of recapitulation, but to most this looks like 'throwing out the baby with the bath water.'"

Carl Sagan not only keeps the baby but also the bath water. Sagan's article in *Parade Magazine*, April 22, 1990, entitled "Is it Possible to be Pro-Life and Pro-Choice?" he shifts back to the myth of embryonic recapitulation. Describing the development of the human embryo, he writes:

"By the third week … it looks a little like a segmented worm … By the end of the fourth week … something like the gill arches of a fish or an amphibian have become conspicuous … it looks something like a newt or a tadpole … By the sixth week … reptilian face … By the end of the seventh week … the face is mammalian, but somewhat pig-like … By the end of the eighth week, the face resembles a primate, but is still quite human."

Are we not shocked that a well-known scientist, in this 1990 presentation for supporting abortion, would resort to the "science fiction of recapitulation"? – A theory that for many years has been debunked by many high-profile scientists.

Sagan, not only has the human embryo start off in a sketchy and incomplete way to indicate that evolution has occurred, but he has it retrace its evolutionary history. That is, climb its alleged family tree through its embryological development.

As *The Revised & Expanded Answers Book* states it:

"This [Sagan's above statement] is straight from Haeckel. A human embryo never looks reptilian or pig-like. A human embryo is always a human embryo, from the moment of conception; it is never anything else, contrary to what Sagan implies!" (Pp.119-120).

It would be in order, at this point, to have a summary of what the creationist camp knows about the history of embryology and how it has led up to the accepted belief in recapitulation. At the beginning of the 19th century, many scientists noted that embryos of widely divergent species seemed to be similar in form and conformed to a basis plan; in the early years of Darwin and his contemporaries, similarities of embryos was believed to determine ancestral relationships. Von Baer, in 1828, claimed that all vertebrates began their embryological lives with gills; Fritz Müller gave Von Baer's theory the recapitulation spin and stated that *evolution of the species is a historical document* displayed in the development of each new individual. Ernst Haeckel was inspired by

Muller's work and eventually came up with the "biogenetic law" or "ontogeny repeats phylogeny." In 1874, Haeckel wrote *The Evolution of Man* which contained fraudulent drawings in order to support his belief in the doctrine of recapitulation.

How does the evolutionary camp respond to this knowledge gained by the creationists? Evolutionists are just as much aware of Haeckel's unsound speculations and inaccurate descriptions as are the creationists. Haeckel's doctored drawings have left some evolutionists feeling ashamed although they had nothing to do with such drawings. Notwithstanding, they say that creationists have misused the information of bad science to discredit all the evolutionary information of good science; that Haeckel's forgeries is "old news" and creationists capitalize on this piece of information to undermine the valid science of embryology.

One can agree that all evolutionary scientists are not responsible for what Haeckel did over a hundred years ago. The doctrine of recapitulation should fall on its lack of scientific merit and not because of what Haeckel did so long ago. Creationists do not need to spend much time in researching the shortcomings of Haeckel. More time should be spent on making it clear to the public why his "biogenetic law" wasn't a valid law in the first place or why the teaching of "ontogeny recapitulates phylogeny" lacks scientific evidence. Later, this appendix will seek to carry out these very suggestions.

Evolutionists are stating that Haeckel's forgeries are irrelevant to the cause of evolution or Darwinian mechanisms. Therefore, should the exposure of Haeckel's deception give creationists cause to let up their assault on evolution and become lenient in meeting the challenges of the Darwinian Theory? Obviously, the "exposure" is no cause for creationists to grow careless. The fact that creationists are simply aware that Haeckel's illustrations were fraudulent and his theory abandoned by 1910 becomes "old news." Evolutionists make the assertion that the very reason the biogenetic law has been rejected is because Darwinian science, by 1910, had conclusively disproved and abandoned Haeckel's theory (This is according to Gould's book *I Have Landed*, P.316). In other words, evolution is a "self-correcting science." Gould makes a good thing out of a bad thing that occurred in the history of an evolutionary scientist. I cannot but help citing the evolutionary maxim which I created for them to meet such situations just cited - "All things work together for good to those who love evolution and are the called according to its purpose."

Before launching an expose on the drawbacks of evolutionary embryology and answering some of the questions posited by evolutionists, a few moments must be taken in defending the creationist stance. To begin with, creationists have not misused information concerning Haeckel's drawings. If anyone should be guilty of misusing information it would be the evolutionist who has persistently used the drawings, while knowing better. Creationists keep the public posted on Haeckel's phony drawings due to the fact, these pictures keep appearing in high school and college text-books and educators are doing nothing about it. It should be noted that not all educators and teachers are guilty of this indictment. There are many who keep up with general scientific history and refuse to teach Haeckel's biogenetic law in their classrooms. True, some creationists go overboard in presenting Haeckel's short comings in their writings but, for the most part, there is good balance in the creationistic literature. A host of scientists who are dedicated to creationism, each year write hundreds of books and articles to inform the public as to the falsity of the Darwinian doctrine which includes the fanciful theories purported by evolutionary embryologists. All such works are written with the goal of reaching scientific conclusions based on intense investigation. The creationist is dedicated to recording the facts that can be observed in nature as opposed to the philosophical dogma which have no observable facts.

We will proceed by briefly tracing the history of embryology, *only this time with the creationist on trial. He, to the best of his ability, will respond to the evolutionists' inquiries.* Let's first review the following fact: at the beginning of the 19th century biologists recognized, in their study of embryonic development, there were similarities in form and external detail (even in diversified forms) and general conformity to a basic plan. Darwin and contemporaries latched on to this so-called fact and translated these findings as evidence that animals evolved from a common ancestor. The similarities were easily detected and more pronounced in the embryo stages than in the adult stage.

(1)*The creationist's task is to explain these similarities*:

In 1942, the embryologist, Doctor Cyril B. Courville, recognized that the similarities in embryologic development on the passage of embryos from the simple to the complex can best be explained by the logic of necessity. Marsh summarizes this postulate:

"There is no other way for two separate and distinct species to develop when starting from a single cell and aiming for an ultimate goal in which similarities still exist, but to follow a somewhat parallel course."

Evolution, Creation, and Science

Frank Lewis Marsh

P.247

Again, it is only to be expected that as the cells of various organisms increase, the differences between the embryos will also increase. When embryos follow the most natural and direct route from a one-celled egg to a many-celled and highly complex adult, the embryos are going through the process of logical necessity.

In *The Revised & Expanded Answer Book*, P.122, it states:

"To construct anything, you begin with something without shape, or with a basic form and then build upon that … … … A fish embryo, however, could never become a human embryo (or vice versa) because a fish embryo has the coded instructions only for making a fish (The coded instructions were not known in Darwin's day)."

The evolutionists, Darwin and his contemporaries, looked at the similarities as they existed in the comparison of animal embryos and concluded – as was mentioned before – these similarities spelled out evolution because they were the result of common ancestry. On the other hand, the creationist looked at the same similarities and concluded that the classic Von Baer's pattern of embryos converging together and then diverging away from each other could not be explained through the process of evolution. Also, the creationists pointed to an intelligent Designer who made living things to follow a similar pattern in their embryonic development. God did this to show that He alone is Creator. The creationists believed that this pattern of similarity did not result from common ancestry.

Once again, history recorded in the realm of science, one's conclusions depended upon one's point of view. The subjective nature of these conclusions made by evolutionists and creationists (back in time) was essentially obvious. However, no longer is this the case. With the evidence to be presented, we should discover why it is more sensible in our day and time to assume the creationist's position. We have spoken to the issue of *similarities* but now let us mention *differences* in embryo development and how these differences pose a problem for evolutionists.

The Revised & Expanded Answers Book, P.122, contains excellent points which can be applied to this issue:

"There are interesting exceptions to Von Baer's laws ... Vertebrate embryos, [in] the stage showing the pharyngeal clefts ... look somewhat similar, but at earlier stages they are quite different!

W.W. Ballard said:

'... From very different eggs the embryos of vertebrates pass through cleavage stages of very different appearance, and then through a period of morphogenetic movements showing patterns of migration and temporary structures unique to each class. All then arrive at a pharyngula stage, which is remarkably uniform throughout the subphylum, consisting of similar organ rudiments similarly arranged (though in some respects deformed in respect to habitat and food supply).'"

Because the earlier stages of embryo development differ, the pattern of similarity could not have resulted, as Darwin contended, from common ancestry. That is, "the differences at the earlier stages gave no support to a naturalistic explanation for similarities at the later pharyngeal stage being due to common descent." [Ibid.P.123]

In a further point, *The Answers Book* directs our attention to Sir Gavin de Beer. He "addressed the problem of the lack of a genetic or embryological basis for homology (a similarity between two organisms due to inheritance of the same feature from a common ancestor) ... in a monograph titled *Homology, an Unsolved Problem.* Although De Beer believed in evolution, he showed that *similarity is often only apparent and is not consistent with common ancestry.*" [Ibid. P.123, Emphasis mine]

Thus, it is more sensible to assume the creationist's position since it is closer to scientific facts. The evolutionist has no basis for his theory of common ancestry; the crux of his position is invalidated by the lack of homogeneity of embryological structures; the similarities are only apparent. What is more, identical organs of closely allied species use cells and tissues from different sources – thus, destroying the homological argument.

So much for the contention of David Quammen who, at the beginning of this Appendix and regarding the similar structures of embryological patterns, directed our attention to the superfluous remarks of Darwin in what Quammen assumed to be the only answer to similarities – "Because, Darwin wrote, 'the embryo is the animal in its less modified state' and that state 'reveals the structure of its progenitor.'" Apparently, Quammen enjoys quoting the wrong authority; he does it so often.

(2) *Another task for the creationist is to explain "gill slits"*:

Karl Ernst von Baer (1792-1876) assumed, in his observations, that he saw gill slits in vertebrate animals. Unfortunately, this observation supplied wood for the fires of evolution. Gill slits in vertebrate animals became another standard case in embryology that evolutionists use as proof for their theory of common descent.

Von Baer did not believe that a human embryo grows gill slits because we evolved from an adult fish (The recapitulatory explanation), but because all vertebrates begin their embryological lives with gills. The fish departs least from this basic condition in their later development, whereas mammals as most "advanced," lose their gills in their departure from the general vertebrate form. But, was this observation of Von Baer really true?

Embryologists know, from modern-day observation, that these "so-called 'gill slits' are really wrinkles in the throat region. This body tissue becomes the palatine tonsils, middle ear canal, parathyroid gland, and thymus in humans."

The Human Body: An Intelligent Design

Alan L. Gillen, Frank J.Sherwin III, Alan C. Knowles

P.32

In the various classes of fish, gill arches allow circulation of water past the gills but the comparable clefts in man do not open. The human embryo never at any time develops gill slits which are claimed by evolutionists to be a part of redundant vestiges of a former evolutionary past. The markings on a human embryo, which superficially look like the gill arches or slits on a fish, never have any breathing function.

In the *Modern Creation Trilogy* (Volume   Two), under the section titled "Evidence from Embryos", there is this significant comment:

"Embryologic researchers have shown conclusively that the embryo never has gill-slits or a tail or any of the supposed evolutionary recapitulations at all. It is programmed – via the genetic code in the DNA molecules – to develop into a human being right from the start, and every stage in its development is essential to reaching that goal."

Henry M Morris

John D.Morris

P.235

British embryologist Lewis Wolpert maintains that a higher animal, like the mammal, passes through an embryonic stage when there are structures that resemble the gill clefts of fish. He writes:

"But this resemblance is illusory and the structures in mammalian embryos only resemble the structures in the *embryonic* fish that will give rise to gills."

Swiss embryologist Gunter Rager explains that pharyngeal pouches which appear in the neck region are purely descriptive ...He writes;

"In man, however, gills do never exist."

[The two embryologists quoted above are in the book "Icons of Evolution" written by Jonathan Wells, P.106]

Wells makes his own comments in this manner:

"The only way to see 'gill-like' structures in human embryos is to read evolution into development.

"To put it bluntly: There is no way 'gill-slits' in human embryos can logically serve as evidence for evolution.

"Gills are not embryonic structures, not even in fish.

"So recapitulation continues to rear its ugly head. Although biologists have known for over a century that it doesn't fit the evidence, and although it was supposedly discarded in the 1920s, recapitulation continues to distort our perceptions of embryos."

Ibid. P. 106

In summary, "gill slits" no longer stand up under the close analysis of embryonic researchers and a "Classic Proof" of evolution turns out to be another myth of the biogenetic law. Once more, for emphasis! The evolutionist, who displays confidence that "gill slits" serve as one of the signs of homology and is indicative of a common ancestor, is countermined by the findings of the true science of embryology. What he thought was proof for his position is not really a "proof" at all.

In returning to the history of embryology, we will consider Fritz Müller (1821-1897). Müller developed von Bear's theory of embryonic growth and differentiation (the development of organs and body parts in ontogeny from simpler antecedent structures). Müller gave it a recapitulation spin and he claimed that the evolution of the species is a "historical document" displayed in new individuals.

On September 10, 1860, Darwin wrote to Asa Grey that embryology is the "strongest class of facts in favor of change of forms." Many creationists assume that Darwin meant recapitulation. Were they correct?

We should remember that Müller's first evolutionary interpretation of recapitulation did not appear until he wrote "Fur Darwin" in 1864. Darwin, in fact, accepted the observations of Von Baer – a denial of recapitulation. However, when Darwin's *Origin* was published in 1859, he did give obvious evolutionary meaning to Von Baer's works. Did Darwin include recapitulation as part of this meaning and at this point in history? Since there is still doubt, we will consider this matter more closely in Appendix lll.

Remembering that Ernst Haeckel (1834-1919) ran with Müller's theory and made this theory into a "biogenetic law" you will remember also, that Haeckel more than any other, advanced the idea that ontogeny repeats phylogeny - the doctrine of recapitulation. Haeckel's deception and dishonesty, previously, have been referred to. However, in the research of Stephen Jay Gould, there is an interesting point that merits mentioning. The late Professor Gould, no matter how one views him, cannot say he did not engage in intensive research for his essays. As a creationist I read all ten volumes of his essays and after reading his 300th essay, concluded through his writings I learned among other things, a great deal about the history of outstanding scientists – creationists as well as evolutionists [I even read *The Hedgehog, the Fox, and the Magister's Pox* that was edited following the death of Steven Jay Gould. In spite of my fellow creationists and their feelings about Gould, I unashamedly shed some tears when I learned of his death. With out being dramatic, I considered him a fellow human being and a fellow sinner in need of God's love and forgiveness]

I often quote from Darwin. The public should know what he had to say – in context – about most of the subjects covered in this work. Also, I often quote Stephen Jay Gould who was not only a strong advocate of natural selection, but made Darwin his hero of science. Years ago, I studied New Testament Greek under a teacher who said, "If you want to know what the New Testament and Paul really teaches, go to the Greek. If you want to know what the Greek really teaches, then go to Phillips." I can almost hear this same teacher saying, "If you want to know what evolution teaches concerning natural selection and survival of the fittest, go to Darwin. If you want to know what Darwin really teaches, go to Gould."   I will now return to Gould's interesting point which involves Haeckel. ... ...

In the early part of the 1980's, Gould was the Alexander Agassiz Professor of zoology, when he found on the open stacks of the Museum's library, Louis Agassiz's personal copy of the first (1868) edition of Haeckel's *The Natural History of Creation* (an indifferent librarianship of a past generation had left this priceless treasure to open access). Into this copy, Agassiz had penciled copious marginal notes. This copy was from Agassiz's library which was passed into the museum's general collection. Some forty pages worth, in typed transcriptions, had been made up from these notes. Gould could not read the scribbling but his secretary, Agnes Plot, could read the archaic German script and transliterated the squiqqles into readable German in Roman type. Gould was able to read this German and in *I Have Landed*, page 315, he states "I could finally sense Agassiz's deep anger and distress."

In 1868, Agassiz at age sixty-one was feeling old and physically broken. Although Agassiz was a special creationist, he set forth some ideas that would not be acceptable with creationists of today. Nevertheless, he did not embrace the new Darwinian model. In fact, Agassiz and Von Baer were the most brilliant opponents of Charles Darwin. Agassiz, all during his lifetime, had a continued opposition to evolution. Gould read the notes that were penciled into Agassiz's copy of Haeckel's *The Natural History of Creation*. [Side note: Haeckel, in placing the word "creation" within his title did not attribute the works of nature to a Creator-God; Haeckel was an atheist who made use of the word "creation" for indicating the mechanical workings of evolution]

Gould wrote in *I Have Landed*, P.315:

"He (Agassiz) immediately recognized what Haeckel had done, and he exploded in fully justified rage. Above the nearly identical pictures of dog and human embryos, Agassiz [realizing that the eye slit, umbilicus, etc, were artistically crafted similarities mixed with inaccuracies] wrote: 'Where were these copied from?'" Gould mentions other deviations in Haeckel's drawings of the dog, chicken, and tortoise embryos. Haeckel made the assertion that "You cannot discover a single difference among them." Agassiz derisively, in his penciled notes responded, "This identity is not true ... these figures were not drawn from nature ... Abscheulich." Agassiz's anger reached the boiling point because of Haeckel's falsified drawings and he threw up his arms in utter disbelief and then wrote "Abscheulich!" ("Atrocious!").

Haeckel had always been a thorn in the side of Agassiz. He did not appreciate Haeckel's materialism, his vicious swipes at religion, and his exaggerations (Gould). We can feel Agassiz's consternation as he came to the realization that Haeckel's book, *The Natural History of Creation*, was a scientific fraud. There was little that escaped Agassiz's attention; he was an expert in geology, paleontology, and zoology; he was also at home with vertebrate embryology – a subject of extensive personal research and writing (Gould's comments on Agassiz).

Agassiz lowered the final curtain on the history of Haeckel's lies and forgeries by pronouncing the justified criticism – Atrocious! Evolutionists as well as creationists can step back from this point in the history of embryology and exclaim, Atrocious!

Let us discuss further, Haeckel's biogenetic law and its scientific distortions. In some evolutionary literature, there are at least five items cited to back up the biogenetic law: the yolk sac, gill slits, tail, the so-called "embryonic six kidneys development" of mammals including the human embryo, and the so-called "two-chambered and three-chambered of the mammalian heart (including humans) in the early embryonic stages."

The reader can recall the coverage given to gill slits in this appendix. Also, he might possibly recall the coverage given to the tail in Chapter 7 where "non-functioning organs" were discussed.

(3)*Since, so far, nothing has been written in this work about the so-called "yolk sac" we will consider a further task of the creationist is to explain this left over (?) organ* :

It should be observed that the remainder of this chapter will not only answer the alleged supporting evidence for evolutionary embryology but, in fact, turn the evidence around and have it stand against Haeckel's biogenetic law. That is, those organs which are considered as "left over" and "non-functional" will be seen plainly to have functions and also necessary to the life of the organism.

Returning to the matter of the "yolk sac", *The Human Body: an Intelligent Design* makes this observation:

"The so-called 'yolk sac' is really a blood sac. Blood cells originate in the structure. Later, bone marrow develops from this tissue in the human fetus."

P.32

The recapitulation terminology "yolk sac" in mammal embryos, including human embryos, obscures the true function of what should be referred to as the blood sac. The so-called "yolk sac" applied to human embryos is said to have no function because it was "left over" from animals who subsisted on stored food in the form of yolk. Since the human embryo does not stay alive in this manner, evolutionists attempt to "feed us" the old recapitulation argument that the "yolk sac" in humans has no function and is, therefore, vestigial. However, the above reference makes it clear that this structure should be called the blood sac since it is here that blood cells originate for the later development of the bone marrow.

An amazing happenstance is how, in the light of all the facts we have amassed regarding the science of embryology, there are phenomena still being interpreted as relics from the past. What was once held by evolutionists to be a strong argument for "non-functioning" organs can no longer become of use in the evolutionary arsenal.

4) *Nevertheless, an argument still being told to the public (in the study of embryology) that it can be ascertained, mammals eliminate four kidneys because they are functionless organs. Therefore, it becomes another task of the creationist to eliminate this concept and place it under the listings of false science:*

Some evolutionists claim that mammals develop six kidneys and end up with two as they eliminate four. The four eliminated kidneys, they say, are explainable only from the standpoint of evolution. They further claim that these four eliminated kidneys are making "an empty gesture to the past by being functionless organs." However, present day authorities and embryologists have already proved that these four kidneys named the pronephros and the mesonephros are not functionless but have a definite function during embryonic life. The authorities claim that these pairs of organs (pronephros, mesonephros, and metanephros) are developed in succession.

Keibel and Mall states:

"Each new excretory organ supplants its predecessor, and the last to develop, the third, the metanephros, becomes the permanent kidney[s]. The pronephros and the mesonephros are merely provisional kidneys, whose activities become superfluous and are supplanted on the formation of a new excretory organ."

Manual of Human Embryology,

Vol. 2

Franz Keibel and F.P. Mall

P.770

Thus, the series of *developmental necessities* are not even close to the concept of recapitulation with its four kidneys making an empty gesture to the past.

5) *The final task of the creationist is to explain the so-called two-chambered and three-chambered of the mammalian heart (including humans) in the early embryonic stages.*

Ballard comments:

"No false biological statement has had a longer or more popular life than the one about the ontogeny of the four-chambered mammalian heart recapitulating its phylogeny from the two-chambered fish heart and the three-chambered amphibian or reptilian heart. Often repeated in the days when Haeckel's so- called biogenetic law was widely and uncritically accepted, it can still be found in some elementary texts, side-by-side with correctly labeled diagrams of the four-chambered fish heart and the five-chambered frog heart."

Comparative Anatomy and Embryology

William Ballard

1964

(Ballard quoted on page 421 in *Creation – Accident or Design?* By Harold G. Coffin)

Gish adds an excellent point:

"If the human embryo recapitulates it's assumed evolutionary ancestory, the human heart should begin with one chamber and then develop successively into two, then three, and finally four chambers. Instead, the human heart begins as a two-chambered organ which fuses to a single chamber, which then develops directly into four chambers. In other words, the sequence is 2-1-4, not 1-2-3-4 as required by the theory."

EVOLUTION: the Fossils STILL say NO!

Duane T. Gish

P. 358

As evolutionary evidence, five points have been covered in embryology to demonstrate the correctness of Haeckel's biogenetic law. These points have been refuted not only by demonstrating the incorrectness of Haeckel's biogenetic law but by placing his law in bold profile as one of those tenets of false science. This appendix also contains other germane issues on the subject of embryology. For examples: an explanation of the apparent similarities in embryological development; the pointing out of differences in embryological development; an examination of the fact there is no basis for homology in embryological structures.

This appendix will close with observations made by two scientists pertaining to the study of embryology. The first observation is made by Stephen Jay Gould – an evolutionist and the second by Duane T. Gish – a creationist.

Gould comments:

"Evolution is strongly constrained by the conservative nature of embryological programs. Nothing in biology is more complex than the production of an adult vertebrate from a single fertilized ovum. Nothing much can be changed very radically without discombobulating the embryo. The order of life, and the persistence of nearly all basic anatomical designs throughout the entire geological history of multicellular animals, records the intricacy and resistance to change of complex development programs, not the perfection of adaptive design in local environments." [Emphasis mine]

"Through a Lens, Darkly"

Natural History (September 1989)

Stephen Jay Gould

For some reason, Gould often quoted from the Bible. He might have done this to prove that he was familiar with the Scriptures used by creationists in their writings; just to poke fun at the authority of the Bible; or he may have made use of the "flavor" of scripture to vaunt his literary skills. The title of Gould's article is taken from the words of the apostle Paul – "For now we see through a glass, darkly … …" (I Corinthians 13:12). What Paul actually said, in the original Greek manuscript was: "What we see now is like the dim image in a mirror …" However, Gould, in his clever way with words, changed the mirror to "lens" signifying either the lens of a fetoscope which can photograph every stage of the human embryo during its development or the lens of a microscope – a main instrument employed in the science of embryology. Fetoscope or microscope will never bring the fetus of any mammal into clear focus so far as providing any useful evolutionary information. Evolutionary embryologists will ever have their vision remain in darkness. Nevertheless, Gould's both shocking and candid statement concerning the reasons for the conservatism of the embryological programs is quite amazing.

In reading Gould's remarks, we may have thought ... ... What a waste of the taxpayer's money for embryological research based on evolution! No wonder the programs are conservative! Here is Gish's apropos remarks:

"Years and years of embryological research was essentially wasted because people, convinced of the theory of evolution and that embryos recapitulate their evolutionary ancestory, spent much of their time in embryological research trying to develop phylogenies based on the data of embryology. As I mentioned earlier, embryologists have abandoned the theory of recapitulation. They don't believe it. They know that it is not true ... it produced bad research rather than the good research that it should have done." (Quote is taken from Luther D. Sunderland's book, *Darwin's Enigma* at page 119). This appendix should have provided its readers with more than enough information to substantiate and agree with Dr. Gish's observation.

APPENDIX III

DARWIN AND THE BIOGENETIC LAW

Darwin is believed (by many creationists) to have made great use of the biogenetic law in *The Origin of Species* and *The Descent of Man*. This appendix will seek to establish, although the biogenetic law was formulated long after Darwin's *Origin* he, nevertheless, made use of <u>recapitulation</u> on which the biogenetic law was based. Haeckel confirmed his biogenetic law in 1876 after being inspired by Fritz Müller's paper of 1864 – both works coming after the 1859 edition of Darwin's *Origin*. How could Darwin have made use of the biogenetic law when it would not be published by Haeckel until seventeen years after the 1859 date?

Rather, didn't Darwin accept Von Baer's embryonic laws of 1828 which were a direct refutation of recapitulation? In fact, Darwin denied none of Von Baer's observations of 1828. Referring to Von Baer, Darwin wrote in his autobiography:

"Hardly any point gave me so much satisfaction when I was at work on the *Origin* as the explanation of the wide difference in many classes between the embryo and the adult animal, and the close resemblance of the embryos within the same class."(This quote from Darwin's autobiography is in *Ontogeny and Phylogeny* by Stephen Jay Gould, P.71).

In other words, Darwin became aware, *while writing his Origin*, that an evolutionary interpretation of great potential was contained within the theory of Von Baer – a fact of which Darwin took advantage. He discovered that Von Baer's studies strongly indicated that the embryos of man and other mammals, because of their resemblance, could be used as a fact in proving the evolutionary theory of a common progenitor.

Gould writes: "De Beer has succinctly summarized Darwin's views in relation both to Von Baer and the recapitulationists in commenting upon this and other passages in the essay of 1842: 'Here, Darwin is following Von Baer in the latter's law of embryonic resemblance ... Darwin's adoption of this view shows that he rejected the transcendental theories of Serres and of Meckel on which Haeckel later based his theory of recapitulation ... Species do not pass through the adult stages of their ancestors during their development, and there is no pressing back into earlier stages of development of characters which first appeared at later stages,'" [Ibid. P.421].

In other words, none of Darwin's arguments (as De Beer makes clear) could be equated with Haeckel's theory of recapitulation.

Darwin explicitly denied recapitulation in his 1842 essay but, in contrast, his book on *Origin of the Species*, 1859, stated:

"For the embryo is the animal in its less modified state; and in so far it reveals the structure of its progenitor" (P. 449).

"As the embryonic state of each species and group of species partially shows us the structure of their less modified ancient progenitors, we can clearly see why ancient and extinct forms of life should resemble the embryos of their descendants – our existing species" (P.381).

In 1859, the above statements presented Darwin's views for tracing of lineages by the similarity of ancestral adults to their own embryos.

As Gould continues in his *Ontogeny and Phylogeny, we need to pay particular attention to his comment following Darwin's above statements* :

"But is not Darwin perilously close to recapitulation at this point? Are we not splitting hairs in attempting to draw a distinction between the actual recapitulation of adult stages and the repetition of embryonic stages that resemble ancestral adults? What difference does it make? Both claims use embryonic stages to trace lineages in the same way" (P.73).

Comparing Darwin's remarks made in his 1842 essay with future remarks in his *Origin* of 1859, has to leave us in grave doubt as to the nature of his convictions. Part of Darwin's theory did not include Haeckel's "biogenetic law" of 1876 but it certainly included recapitulation which was wide spread at the time when Darwin wrote his 1859 *The Origin of Species*. Darwin does seem to have incorporated recapitulation as part of his evolutionary doctrine before Haeckel made the same theory, years later, into a law.

David Quammen, at the beginning of Appendix II quotes Charles Darwin's words, "The embryo is the animal in its less modified state [and that state] reveals the structure of its progenitor."

Was Darwin, as Gould questions, "perilously close to recapitulation at this point?" I believe he was not only perilously close but, in fact, was espoused to the doctrine of recapitulation in the year 1859 and which was the basis of the biogenetic law in the year 1876. Therefore, Darwin in anticipation of Haeckel's law was already writing about recapitulation (the crux of Haeckel' law) in *The Origin of Species*.

The obvious conclusion as far as the biogenetic law is concerned Darwin, in a sense, "made great use of it in *The Origin of Species* and *The Descent of Man*."

[Quote from Henry M. Morris's Impact,

September 1988]

Once again, in Quammen's attempt to present "science" to the public with his fervid quote of Darwin, he merely presented the antiquated and obsolete doctrine of recapitulation.

ACKNOWEDGEMENTS

Utilizing a word processor for the first time in my life, can be a frustrating process. I am close to seventy years old and not prone to try anything new. When I informed my son, Nick Walden, I wanted to write a book by using a typewriter, he was truly gracious – he merely snickered and didn't laugh out loud. When Nick inquired did I ever use a computer, I responded by saying that I would never be capable of handling one. But thanks to him, I learned the essential steps.

I brought a beginners manual on how to operate a computer. I will not say that I did not acquire some helpful information but I must have telephoned Nick a thousand times and asked him to drive over to my home and help me out of a technical predicament. I still marvel as I think back on his ability to fly over the keyboard with his lightning quick hands. Nick, I thank you for never complaining when I called you away from your busy work schedule! – at least not to my face. I also thank you for taking care of the needful steps in the publishing management. I also want to acknowledge my other sons for their support in whatever I have attempted to do in my later life. The Pittack order of life's events was Brian, David, Daniel, and Kyle. Thank you, Kyle, for not sending me to bed when you walked by my computer table at three o'clock in the morning! I also thank you for performing your magical computer tricks when I sat in a trance before my machine – not knowing what to do next.

I also would like to acknowledge my lawyer friend, Mike Pincher, who heard about my being a creationist and has been visiting with me for the past four years. Mike is an ardent creationist whose sincerity and intelligence level makes for a coveted friendship. I thank you, Mike, for listening to and putting up with the broken speech that followed my stroke! I am grateful to you for volunteering to be my chauffer and for driving your own car to those meetings on creationism in out-of-the-way places; also seeing that I got to local meetings on Creation Science in Lancaster, California. Mike, I am eager to read the book you are now preparing! If it's anything like your conversations, then we are all in for a treat.

I thank my oldest son, Brian Pittack, for using his eagle eyes in the editing of this book.

Finally, I would like to thank my wife, Kathleen, who has been with me for almost three decades. She gave me abundant support in those fearsome months when I protracted cancer. I couldn't have gotten through my operation without Kathleen and God. I thank you, Kathleen, for loving me through the bad times as well as the good times; Thank you, God, for loving me through any kind of times.

SOURCES

BOOKS (Single Author)

Austin, Steven A. (Editor), *Grand Canyon* (Monument to Catastrophe), Institute for Creation Research, Santee, California, Copyright © 1994.

Ballard, William *Comparative Anatomy and Embryology*, 1964 (Ballard is quoted on page 421 in *Creation – Accident or Design?* by Harold Coffin).

Behe, Michael J., *Darwin's Black Box*, Simon & Schuster Inc. First Touchstone Edition, 1998. Copyright © 1996 by Michel J. Behe.

Clark, Austin H. *The New Evolution:Zoogenesis*, 1931

Clark, Harold W., *Fossils, Flood, and Fire*. OUTDOOR PICTURES, Escondido, California, Copyright, 1968, Harold W. Clark.

Clark, Harold W., *New Creationism*, Southern Publishing Association, Nashville, Tennessee, 1980 Copyright © 1980 by Southern Publishing Association.

Clark, Harold W., *The Battle Over Genesis*, Review and Herald Publishing Association, Washington, D.C., 1977. Copyright © 1977 by Review and Herald Publishing Association.

Coffin, Harold, *Creation – Accident or Design?* Review and Herald Publishing Association, Washington, D.C., 1969. Copyright © 1969 by the Review and Herald publishing Association.

Darwin, Charles, *The Origin of Species*, (1859) and the *Descent of Man* (1871). Bennett A. Cerf. Donald S. Klopfer, The Modern Library, Random House, Inc., New York.

Dawkins, Richard, *The Blind Watchmaker*, W. W. Norton, New York, 1987.

De Beer, Gavin, *Homology, an Unsolved Problem* (Oxford Biology Reader, Oxford University Press, 1971).

Denton, Michael, *Evolution: A Theory in Crisis*, Adler & Adler, Publishers, Inc., Chevy Chase, MD, 1985. Copyright © 1985 by Michael Denton.

Gish, Duane T., *Creation Scientists Answer Their Critics*, Institute for Creation Research, El Cajon,CA, First Edition, 1993. Copyright © 1993.

Gish, Duane T., *The Amazing Story Of Creation*, Institute for Creation Research, El Cajon, CA, 1990. Copyright © 1990 Institute for Creation Research.

Gish, Duane T., *Evolution:The Fossils Still Say No!* Institute for Creation Research, El Cajon, California, First Edition Revision, Copyright © 1995.

Gould, Stephen Jay, *Bully for Brontosaurus*, W.W. Norton & Company, Inc., New York, N.Y., 1991. Copyright © 1991 by Stephen Jay Gould.

Gould, Stephen Jay, *Dinosaur In A Haystack*, Harmony Books, a division of Crown Publishers, Inc., New York, New York. Copyright © 1995 by Stephen Jay Gould.

Gould, Stephen Jay, *I Have Landed*, Harmony Books, New York, New York, Copyright © 2002 by Turbo, Inc.

Gould, Stephen Jay, *Leonardo's Mountain of Clams and the Diet of Worms,* Three Rivers Press is a registered trade mark of Random House, Inc, New York, New York. Copyright © 1998 by Turbo, Inc.

Gould, Stephen Jay, *Ontogeny and Phylogeny*, The Belknap Press of Harvard University Press, Cambridge, Massachusetts. Copyright © 1977 by The President and Fellows of Harvard College.

Gould, Stephen Jay, *The Flamingo's Smile Reflections In Natural History*, W.W. Norton & Company. Copyright © by Stephen Jay Gould.

Gould, Stephen Jay, *The Panda's Thumb*, W.W. Norton & Company, Inc., New York, N.Y., 1980. Copyright © 1980 by Stephen Jay Gould.

Gould, Stephen Jay, *The Structure of Evolutionary Theory*, The Belknap Press of Harvard University Press Cambridge Massachusetts and London, England, 2002. Copyright © 2002 by the President and Fellows of Harvard College.

Hager, Michael W., *Fossils of Wyoming*, Laramie, Wyoming, December 1970, P.44. Bulletin 54 of the Wyoming Geological Survey.

Hanegraaff, Hank, *The Farce of Evolution*, World Publishing, Nashville, TN, 1998. Copyright © 1998 by Hank Hanegraaff.

Johnson, Phillip E., *Darwin On Trial*, Inter Varsity Press, Downers Grove, Illinois, 2nd edition, 1993. Copyright by Regnery Gateway, Inc.

Kerkut, G.A., *Implications of Evolution*. New York: Pergamon Press, 1960.

MacFadden, Bruce J., *Fossil Horses*. Cambridge University Press, 1992.

Marsh, Frank Lewis, *Evolution, Creation, and Science*. Review and Herald, Washington, D.C.; Second Edition Revised, 1947. Copyright, 1944, 1947 by the Review and Herald Publishing Association.

In the above mentioned book, on page 533, there is a quote taken from *Animal Biology*, 1941, by M.F. Guyer.

Marsh, Frank Lewis, *Life, Man, and Time*, Outdoor Pictures, Escondido, California, Revised edition, 1967. Copyright, 1967, by Frank Lewis Marsh.

Mound, Laurence, *Insect*, First American edition, Alfred A. Knopf. Copyright © 1990, Eyewitness Books.

Scripture quotations used in this book are from the Holy Bible, New International Version (NIV), Pocket-Size Edition, Broadman & Holman Publishers Nashville, Tennessee, copyright © 1973, 1978, 1984 International Bible Society.

Perloff, James, *Tornado in a Junkyard*, Refuge Books, Arlington, MA, third printing. July 2000, Copyright © by James Perloff.

In the above mentioned book, on page 27, there is a quote taken from *Not By Chance!: Shattering the Modern Theory of Evolution*, 1997, by Lee Spetner.

Ritland, Richard M., *A Search for Meaning in Nature*, Pacific Press Publishing Association, Mountain View, California, 1970. Copyright © 1970, by Pacific Press Publishing Association.

In the above mentioned book, on page 142, there if a quote taken from *The Evolution of Flowering Plants* by Daniel Axelrod.

Sunderland, Luther D., *Darwin's Enigma*. Master Book Publishers, Santee, California, Second Edition, 1984. Copyright © 1984, Luther D. Sunderland.

Excerpt from a debate between Dr. Duane Gish and Ashley Montague on April 12, 1980, at Princeton University is recorded on page 119.

Webster's Family Encyclopedia, Volume 4, 1995 Edition (this book is not published by the original publishers of WEBSTERS DICTIONARY, or by their successors), Archer Worldwide Inc., Great Neck, New York, USA. Copyright © 1981, 1983, 1986, 1987, 1988, 1989, 1990, 1992, 1993, 1994, Market House Books Aylesbury.

Wells, Jonathan, *Icons of Evolution Science or Myth*, First paperback edition 2002, published in the United States by Regnery Publishing, Inc. An Eagle Publishing Company, Washington, DC 2001, Copyright © 2000 by Jonathan Wells.

Zuckerman, Solly, *Beyond the Ivory Tower*, New York: Taplinger Publishing Company, 1971.

BOOKS (More Than One Author):

Ankerberg, John & Weldon, John, *Darwin's Leap of Faith*, Harvest House Publishers, Eugene, Oregon, 1998. Copyright © 1998 by John Ankerberg and John Weldon.

Behler, John L.; King, F. Wayne, *The Audubon Society Field Guide to North American Reptiles and Amphibians*, a Borzoi Book published by Alfred A. Knopf, Inc., New York. Copyright 1979 under the International Union for the protection of literary and artistic works (Berne).

Davis, Percival; Kenyon, Dean H.; Thaxton, Charles B. (Academic Editor), Of *Pandas and People*, Haughton Publishing Company, Dallas, Texas, Fifth Printing 2004. Copyright 1989, 1993 by Foundation for Thought and Ethics.

Gillen, Alan L; Sherwin lll, Frank J. Knowles, Alan. *The Human Body: An Intelligent Design*, Creation Research Society Books, and Creation Research Society Monograph Series: No.8, Copyright 2001 Creation Research Society, SECOND EDITION. (Statement of Dr. E. Blechschmidt quoted in *The Human Body: An Intelligent Design*, p.33).

Han, Ken; Sarfati, Johnathan; Wieland, Carl; edited by Ballen, Don, *The Revised & Expanded Answers Book*, Master Book, Thirtieth printing: April, 2004. Copyright © 1990 by Creation Science Foundation.

Hickman, Cleveland; Roberts Larry; Larson, Allan, *Integrated Principles of Zoology*, Dubuque, IA; Wm. C. Brown, 1997.

Keibel, Franz; Mall F.P., *Manual of Human Embryology*. Philadelphia: J.B. Lippincott Co., vol.2, 1912.

Leakey, Richard E. & Lewin, Roger, *Origins*, published by E.P. Dutton, a division of NAL Penguin Inc., New York, N.Y., Second Dutton (unillustrated) paperback edition, 1982. Copyright © 1977 by Richard Leakey and Roger Lewin.

Morris, Henry M.; Morris, John D., *Modern Creation Trilogy* (volume two) Science and Creation, Master Books, second printing, November, 1997. Copyright © 1996 by Master Books.

ARTICLES:

Ashton , John F. (editor), *In Six Days* – Why fifty scientists, choose to believe in creation, Master Books, Inc., fourth printing, 2003. Copyright © 2000 by John F. Ashton. In this book quotes are taken from the following three articles:

1)   James S. Alan, "Genetics," Pp.130-131.

2)   Henry Zuill's, "Biology," P.71.

3)   Andrew McIntosh's, "Mathematics," Pp.167-168.

Ballard, W.W., "Problems of Gastrulation: Real and Verbal." *Bioscience* 26(1):36-39, 1976.

Gould, Stephen Jay, *Natural History* (September 1989). "Through a Lens, Darkly."

Lewin, Roger, *Science*, vol.210 (November 21, 1980), Pp.883-884. "Evolutionary Theory Under Fire."

Morris, Henry M., "The Heritage of the Recapitulation Theory," *Impact* 183 (September 1988):1

National Geographic, March, 1992, "A Curious Kingship: Apes and Humans." Article written by Eugene Linden, Photographs by Michael Nichols.

National Geographic, November, 2004, "Was Darwin Wrong? – NO," article written by David Quammen, Photographs by Robert Clark. My entire book is based on a response to the issues raised in Quammen's article.

Perlas, Nicky, *Towards,* vol.2 (Spring 1982), "Neo-Darwinism Challenged at AAAS Annual Meeting."

Pilbeam, David, "Rearranging Our Family Tree," *Human Nature* (June 1978).

Poirier, Jules; Cumming, Kenneth B., "Design Features of the Monarch Butterfly Life Cycle," *Impact* 237 (March 1993):2-3.

Sagan, Carl, *Parade Magazine*, article entitled "Is It Possible to Be Pro-Life and Pro-Choice?" April 22, 1990.

Schwabe, Christian, "On The Validity Of Molecular Evolution," *Trends In Biochemical Sciences* (July 1986).

THESIS:

Barnhart, Walter R., Thesis for Master of Science (El Cajon, California; Institute for Creation Research Graduate School, 1987, Pp.148-150). Reference in *EVOLUTION: The Fossils STILL say NO!* by Duane T. Gish.

TEACHING QUARTERLY:

Younker, Randall W., God's Creation, *Sabbath School*, Personal Ministries Dept.; Silver Spring, MD, July-Sept, 1999.

www.ingramcontent.com/pod-product-compliance
Lightning Source LLC
Chambersburg PA
CBHW021232090426
42740CB00006B/499